HIGHER

MATHS

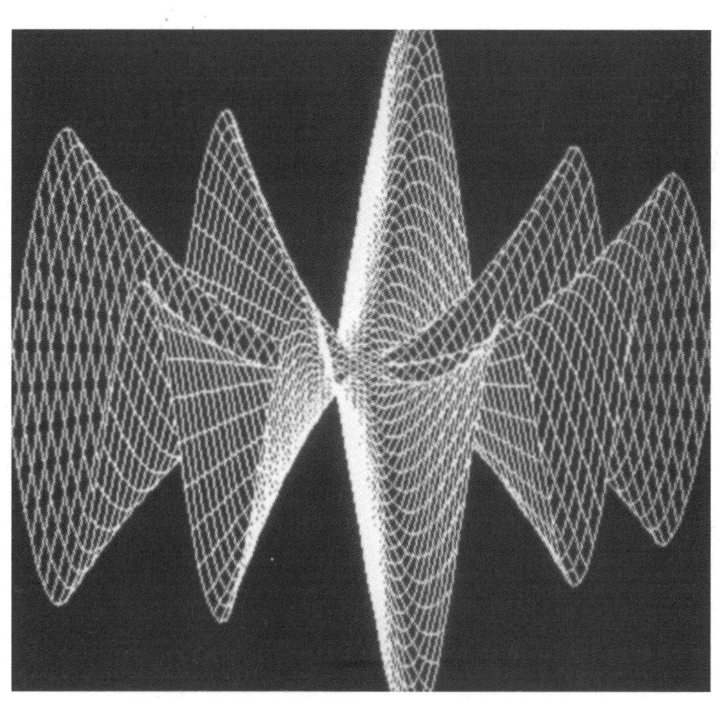

Peter Westwood

with additional material by John Winter

Hodder Gibson

A MEMBER OF THE HODDER HEADLINE GROUP

Acknowledgements

It is my pleasure to record my gratitude to George Grant and Eric McCallum for their helpful comments on, and constructive criticism of the manuscript, the former for a very detailed analysis.

P. W. Westwood

August 2004

Publisher's Note

This book is dedicated to the memory of Peter Westwood, a wonderful colleague and mathematician, who died as this book was in production. A large debt of gratitude is also owed to George Grant, long-time friend and colleague of Peter's, who kindly assisted in the final stages of the book's production.

Thanks also to John Winter of Strathclyde University for providing additional material to update the contents from 2008 onwards.

Orders: please contact Bookpoint Ltd, 130 Milton Park, Abingdon, Oxon OX14 4SB. Telephone: (44) 01235. 827720. Fax: (44) 01235 400454. Lines are open from 9.00–5.00, Monday to Saturday, with a 24-hour message answering service. Visit our website at www.hoddereducation.co.uk. Hodder Gibson can be contacted direct on: Tel: 0141 848 1609; Fax: 0141 889 6315; email: hoddergibson@hodder.co.uk

First published in 2005 by
Hodder Gibson, an Hachette-Livre UK Company
2a Christie Street
Paisley PA1 1NB

Impression number 10 9 8 7 6 5 4
Year 2010 2009 2008 2007

Cover photo shows a 3-D plot of a complex maths function (Science Photo Library)

Typeset in 9.5/12.5 Frutiger by Tech-Set Ltd, Gateshead

Printed and bound in Great Britain by Martins The Printers, Berwick-upon-Tweed

A catalogue record for this title is available from the British Library.

ISBN 13: 9780 340 885529

CONTENTS

Chapter 1	An Introduction to Higher Mathematics	1
Chapter 2	Revision topics	5
Chapter 3	The straight line	10
Chapter 4	Functions and graphs	18
Chapter 5	Differentiation	28
Chapter 6	Recurrence relations	36
Chapter 7	Identities and radians	38
Chapter 8	Quadratic theory	46
Chapter 9	The remainder theorem	50
Chapter 10	Integration	56
Chapter 11	Compound and multiple angles	60
Chapter 12	The circle	63
Chapter 13	Vectors	68
Chapter 14	Further calculus	78
Chapter 15	Logarithms	81
Chapter 16	The auxiliary angle	87
Chapter 17	Typical shorter questions	92
Chapter 18	Typical longer questions	108
Chapter 19	Multiple choice questions	127

(Chapters 3–7 cover Unit 1 of the Higher Mathematics syllabus. Chapters 8–12 cover Unit 2, and Chapters 13–16 cover Unit 3.)

Appendix 1: Examination Technique 135

Appendix 2: Notation 139

Appendix 3: Mathematical Vocabulary and Index 140

Updated sections – important note

The book you are holding is from the 4th printing (or later) of this title. An extra chapter containing multiple choice questions has been added to ensure the contents are appropriate for Higher Maths from 2008 onwards, providing tips and practice questions for the new objective tests section in Paper 1.

Chapter 1

AN INTRODUCTION TO HIGHER MATHEMATICS

Welcome to this Revision Book!

I am glad to hear that you have opted to try to pass Higher Mathematics this session. You are not alone. There will be almost 20 000 other people sitting the exam at the same time as you. You should find that encouraging; if most of **them** can pass, then why not **you**.

I hope you find the course varied and interesting. All of the content is applicable to somebody's future study. You may have to study some parts that you may never meet again, but that is because Mathematics pervades so many other diverse areas of knowledge. The course should help you to think logically and clearly and give you practice in communicating what you have worked out. These skills are useful in all walks of life.

I have addressed this book to pupils in schools, who make up the majority of candidates (candidates is SQA-speak for anyone, school pupil, college student or adult, who sits the exam), but the guidance applies equally to anyone sitting the exam. It is my hope that this book will be of benefit to candidates of all levels, including those who are needing all the help they can get to achieve a pass to those who are aiming to score as close as they can to 100%.

How to use this Revision Book

This is not a course book. You cannot pass the examination solely by studying this book. Do not try to take in all of it at once.
Read small parts at a time and try to digest them. Re-read them occasionally. In this book, I:

1. include hints on how to work throughout the year
2. include hints on how to revise for tests and exams and on exam technique to apply during them
3. itemise the necessary theory and basic knowledge and provide explanatory worked examples, some of which merit further mention under the headings of 4 and 5 below and are listed appropriately on the contents page.
4. help you to recognise certain shorter standard easy questions to build up your mark
5. provide some longer exam standard questions with guidance on how to set about them. You should try these for yourself before studying my solution and comments.

I have always maintained, as did **my** teachers before me, that you cannot learn Mathematics simply by reading. Mathematical knowledge does not rise up off the page and enter your brain. Effective revision almost certainly involves writing, which may be summarising or working through examples. Learning mathematical skills is an activity in which you actually have to become involved yourself before it registers in your mind.

The Higher Mathematics Exam

The exam is set by the Scottish Qualifications Authority, which is the same body that set your Standard Grade or Intermediate exam. These examinations are very fair to candidates. The Higher Maths exam is set by a team, then reviewed by moderators and finally overseen by a scrutineer, who all take great care to ensure that the exam is of the correct standard, that there are no misleading questions, and that the level of difficulty will increase gradually throughout the paper. Hence you may expect that all the language used in the questions should be straightforward, that every question will be related to the syllabus, and all the mathematics that you require to answer the questions will have been seen before in your classroom.

S.Q.A. have published an itemised list of the content that you are required to know for the exam at www.sqa.org.uk/files/nq/C10012.pdf, but this book is also intended to cover all the essential knowledge required, hopefully in a more pupil friendly fashion and with direct hints. In this S.Q.A. list you can see types of question which they regard as being appropriate for pupils who gain a C grade pass. These constitute about two thirds of the marks in the final exam. The other third of the marks is made up of examples which S.Q.A. consider to be more challenging, which are also indicated in their listing. This is taken into account when the order of the questions is decided. This, however, is not an exact science and what S.Q.A. consider challenging, you may find routine, or vice versa, so, in the exam, you should try every question.

The content of the syllabus is divided into three units and you have to pass the unit test for each one, but these should not be great hurdles if you are keeping up with the work.

There is some algebra, geometry, trigonometry and calculus in each unit. Calculus is a new topic which was not included in the Credit syllabus, but you must also retain skills that you learned in Credit, especially factorising, recognising quadratic equations, operations with fractions, using indices, and sketching parabolae and exponential functions.

Working throughout the year

The first thing to realise is that Higher Maths is so much more demanding than credit, and you need to address this from the very start. I do not say this to put you off, but to prepare you for what is to come. The difficulties are not insurmountable because thousands pass every year. You must appreciate the workload that you have taken on and not squander any time. On analysing my school's results every year, it is not always the brightest pupils who do best. Many less outstanding pupils score commendably because they have wanted to pass the exam and have put in the effort to realise their ambition. I have taught many excellent and intelligent pupils who have done really well, but there are almost as many very capable pupils who have failed to secure an A pass. Motivation often counts for as much as ability.

Olympic athletes who win medals do not do so by deciding a month before the Olympics that it might be a good idea to race. They train regularly for years beforehand and schedule their later training towards the games. Similarly, you will not pass Higher Mathematics by deciding to start working for it a month beforehand. The years of practice you have already spent in primary school and in early secondary school are all part of the build-up for gaining a pass in Higher Maths. Your level of application during the year before you sit Higher Maths will have the greatest effect on your result. Another analogy which is useful here could be within your own experience. Nobody learns to play the piano in a month or by practising solely for the half hour preceding their weekly lesson. A small amount often and regularly is the recipe for success. And so it is with mathematics. You need to be geared up all year for the exam at the end of it. **It is the cumulative effect of all your daily efforts that go to producing a good performance in the actual exam.** So it is very worthwhile developing good work habits as a foundation for your learning.

Attendance at class is the most vital ingredient of any successful mathematical education. Other interests do conflict from time to time, e.g. musical, dramatic, or sporting events. There are also medical appointments and possibly even illness to contend with. It is essential that you make up for what you have missed as soon as possible after an absence. Mathematics is a sequential subject in that what you learn one day often forms the basis of what you learn the next. Missing any lesson means that you will also have missed the class discussion which was involved in the introduction to the lesson, which is all part of your learning process. Your teacher will not have a lot of time in class to reconstruct this part for you. Your teacher may be able to devote some time to help you to catch up on a missed lesson but if you can have things explained to you by another pupil, then that actually helps the other pupil to understand the lesson better as well. Equally you can gain confidence through explaining something to someone else who has been absent. So keep up with the work of the class. If you get behind, the gap is more likely to widen than to close.

Keep your work well organised with each topic kept separate, whether in jotters or on loose leaf paper. On the other hand, always remember that this is an artificial division of knowledge for the purposes of teaching and learning. Mathematics is greater than the whole of all these parts, which are relentlessly interwoven.

You cannot achieve a sufficiently high standard of mathematical knowledge and ability or assimilate all the necessary skills in the 40 hours nominally allocated to each unit, so that doing regular homework is an essential feature of the course. Your having bought this book indicates that you already accept the value of homework. It helps to cement each day's knowledge or skill in your mind. If you leave it too late to return to any day's lesson, the effort you have put in in class can be lost. There is also seldom enough time in class to include sufficient practice of skills learnt, so this must be built up at home. Only you can assess the amount of homework that you require to do. It will depend, for example, on how many and which subjects you are studying, how fast you can work, how fast you assimilate new learning. This is one of the main areas where you **take responsibility for your own learning**. This is not just a catch phrase; it is at the heart of being well motivated.

Some pupils see a conflict between homework and part time employment. There should not be conflict; a sensible balance is the answer. It is good for pupils to obtain some work experience and to earn their own pocket money, but the amount of paid work undertaken has to be thoughtfully limited. What you get on your certificate at the end of the year is far more important than whether you can afford to take a Mediterranean holiday with your cronies next summer. If I apply for another job tomorrow, I will still be asked to state what Highers I passed at school. You may need to make some sacrifices to secure the qualifications you require for the job you want.

Your teacher will most probably issue notes, duplicated or otherwise. You should also create your own summary notes, of anything that needs to be committed to memory. Writing them down helps you to learn them initially, and looking over them regularly aids retention. You can highlight appropriate parts of your own handwriting and draw boxes around important bits. It is very useful to make a list of all the formulae that you need to know, adding to the list throughout the year. Keep all the "=" signs in line and lay another sheet of paper over the right hand sides of the formulae, then try to write these down from memory. This process should be done every evening with any new formulae until they are thoroughly known, and perhaps weekly thereafter to keep all the formulae fresh in your mind. Do not forget to learn the formulae backwards as well, e.g. $2\sin A\cos A = \sin 2A$.

Remember that getting stuck is part of learning. It is how you get unstuck that helps you to understand and remember things. Do not be dismayed if your teacher does not simply tell you the answer to any question you may have. You learn more when your teacher can draw the answer from you by judicious questioning, so you have to play your part in this process too.

Your final exam consists of doing a Paper 1 with 24 examples (20 objective, 4 short and extended response) without a calculator, and Paper 2 with a calculator. The more examples you tackle throughout year, the more chance you have of doing well in the final exam. There is an essential place for listening to your teacher in your maths class but you may learn relatively little by that process alone. You **learn** mathematics when you do it for yourself and when you discuss it with others (colleagues and teachers). The seeds of learning are sown in the classroom, but ensure that they do not fall on stony ground.

Chapter 2

REVISION TOPICS

Revision before you start on the Higher course

You may find this section useful if you are just starting the course. It may also be useful later if your lack of knowledge of previous work has already let you down with some of the Higher content.

What You Should Know

2·1 how to solve simultaneously two linear equations in two variables

2·2 the laws of indices

2·3 how to express a surd in its simplest form or with a rational denominator

2·4 how to factorise
- a) using a common factor e.g. $\sin x \cos x - \sin^2 x$
- b) a difference of two squares e.g. $16 - \sin^2 3x$
- c) a quadratic expression e.g. $2\sin^2 x - 3\sin x + 1$

2·5 how to solve quadratic equations (the highest power of the variable is 2)

2·6 how to sketch quadratic and exponential functions (e.g. $y = x^2 + 4x + 3$, $y = 2^x$)

2·7 how to add, subtract, multiply, and divide algebraic fractions

Notes on items 2·1 to 2·7: Revision Topics

2·1 how to solve simultaneously two linear equations in two variables.

Example

Solve for x and y

$$2x + 5y + 14 = 0$$
$$3x - y = 13.$$

(Solution) Method 1 (combinations)

arrange to get numerically the same y-coefficient in both equations [we could do this for x instead, but that needs two multiplications] so multiply the second equation by 5

$$2x + 5y = -14$$
$$15x - 5y = 65$$

adding $17x = 51$ $\Rightarrow x = 3$

substituting $x = 3$ in the second equation $\Rightarrow 3 \times 3 - y = 13$

$\Rightarrow y = 3 \times 3 - 13 = -4$

hence $x = 3$, $y = -4$

Example *continued* ➢

Example continued

Method 2 (substitution)
this is most useful when one of the four coefficients is ± 1;
re-arrange the second equation as $y = 3x - 13$
take great care with this re-arrangement; take more than one line if necessary as this is the most common place for making an error
then substitute into the other equation, $2x + 5y + 14 = 0$, which becomes

$$2x + 5(3x - 13) + 14 = 0$$
$$\Rightarrow \quad 17x - 65 + 14 = 0$$
$$\Rightarrow \quad 17x = 51$$
$$\Rightarrow \quad x = 3$$

now use your original re-arrangement, $y = 3x - 13$, to find the y value easily
$\Rightarrow \quad y = 3x - 13 = 3 \times 3 - 13 = -4$ hence $x = 3$, $y = -4$ as before

2·2 the laws of indices

Key Points

The laws of indices are

I $\quad x^p \times x^q = x^{p+q}$

II $\quad x^p \div x^q = x^{p-q}$

III $\quad (x^p)^q = x^{pq}$

IV $\quad x^p \times y^p = (xy)^p$

also learn $x^0 = 1$, $x^{-n} = \dfrac{1}{x^n}$, $x^{\frac{1}{n}} = \sqrt[n]{x}$, $x^{\frac{p}{q}} = \sqrt[q]{x^p} = (\sqrt[q]{x})^p$ i.e. p is an index, not a multiplier.

Example

Express $\dfrac{x^3 + 2x + 3}{\sqrt{x}}$ in the form $ax^p + bx^q + cx^r$.

(Solution)
$$\frac{x^3 + 2x + 3}{\sqrt{x}} = \frac{x^3 + 2x + 3}{x^{\frac{1}{2}}} = \frac{x^3}{x^{\frac{1}{2}}} + \frac{2x}{x^{\frac{1}{2}}} + \frac{3}{x^{\frac{1}{2}}}$$
$$= x^{(3-\frac{1}{2})} + 2x^{(1-\frac{1}{2})} + 3x^{-\frac{1}{2}}$$
$$= x^{\frac{5}{2}} + 2x^{\frac{1}{2}} + 3x^{-\frac{1}{2}}$$

2·3 how to express a surd in its simplest form or with a rational denominator

Example

a) Express $\sqrt{80}$ in its simplest form.

b) Express $\dfrac{3}{\sqrt{45}}$ with a rational denominator in its simplest form.

(Solution) a) first find the largest perfect square which is a factor of 80
4 yes, 9 no, 16 yes, 25 no, 36 no, 49 no, 64 no, so 16
$80 = 16 \times 5$ so $\sqrt{80} = \sqrt{16 \times 5} = \sqrt{16} \times \sqrt{5} = 4\sqrt{5}$

b) the largest perfect square which goes into 45 is 9,
so $\sqrt{45} = \sqrt{9 \times 5} = \sqrt{9} \times \sqrt{5} = 3\sqrt{5}$

hence $\dfrac{3}{\sqrt{45}} = \dfrac{3}{3\sqrt{5}} = \dfrac{1}{\sqrt{5}}$

if we multiply the bottom line by $\sqrt{5}$ it becomes 5, which is rational, but to keep the fraction the same size we must also multiply the top line by $\sqrt{5}$, i.e. multiply top and bottom by $\sqrt{5}$ (so we are multiplying by 1)

hence $\dfrac{1}{\sqrt{5}} = \dfrac{1}{\sqrt{5}} \times \dfrac{\sqrt{5}}{\sqrt{5}} = \dfrac{\sqrt{5}}{5}$

Note: Don't do silly things like leaving answers in terms of $\sqrt{1}$ or $\sqrt{4}$

2·4 how to factorise a) using a common factor b) a difference of two squares c) a quadratic expression

Example

Factorise a) $\sin x \cos x - \sin^2 x$
b) $16 - \sin^2 3x$
c) $2\sin^2 x - 3\sin x + 1$.

(Solution) a) the common factor is $\sin x$: $\sin x \cos x - \sin^2 x = \sin x\,(\cos x - \sin x)$

b) $16 - \sin^2 3x = (4)^2 - (\sin 3x)^2$
$= (4 - \sin 3x)(4 + \sin 3x)$

c) the factors of $2\sin^2 x$ can only be $(2\sin x)$ and $(\sin x)$, and the factors of 1 can only be 1 and 1
hence $2\sin^2 x - 3\sin x + 1 = (2\sin x \ldots 1)(\sin x \ldots 1)$
the signs are the same (because of the + in front of the 1) they are both − (because of the − in front of the 3 sin x)
$\therefore 2\sin^2 x - 3\sin x + 1 = (2\sin x - 1)(\sin x - 1)$

[Check all the answers mentally by expanding the brackets.]

2·5 how to solve quadratic equations (the highest power of the variable is 2)

Example

Solve a) $2x^2 + x - 1 = 0$ b) $x^2 - 16 = 0$ c) $x^2 - 6x = 0.$

[Note a) three terms b) no x term c) no constant term]

(Solution) a) by factorising

$(2x - 1)(x + 1) = 0$

$\Rightarrow x = \frac{1}{2}, -1$

by the quadratic formula

$x = \frac{-b \pm \sqrt{b^2 - 4ac}}{2a} = \frac{-1 \pm \sqrt{1 - 4 \times 2(-1)}}{4}$

$x = \frac{-1 \pm 3}{4} = -1, \frac{1}{2}$

b) by factorising

$(x - 4)(x + 4) = 0$

$\Rightarrow x = 4, -4$

by taking the square root of both sides

$x^2 = 16$

$\Rightarrow x = \pm 4$

remember to include $\pm$ whenever you take the square root of both sides of an equation

c) This is an easy type of question but it is a favourite for pupils forgetting what to do with it in the exam. Use the common factor

i.e. $x(x - 6) = 0 \Rightarrow x = 0, 6$

2·6 how to sketch quadratic and exponential functions

Example

Sketch the graphs with equations a) $y = x^2 + 4x + 3$

b) $y = 2^x$

(Solution) a) Recognise $y = x^2 + 4x + 3$ as the equation of a parabola

for crossing the y-axis, $x = 0 \Rightarrow y = 3 \Rightarrow (0, 3)$

for crossing the x-axis, $y = 0 \Rightarrow y = x^2 + 4x + 3 = (x + 1)(x + 3) = 0$

$\Rightarrow x = -1, -3 \Rightarrow (-1, 0)$ & $(-3, 0)$

the axis of symmetry is half way between roots, i.e. $x = -2$

$x = -2 \Rightarrow y = -1 \Rightarrow$ turning point at $(-2, -1)$

this is a minimum because the coefficient of x^2 is positive, hence:

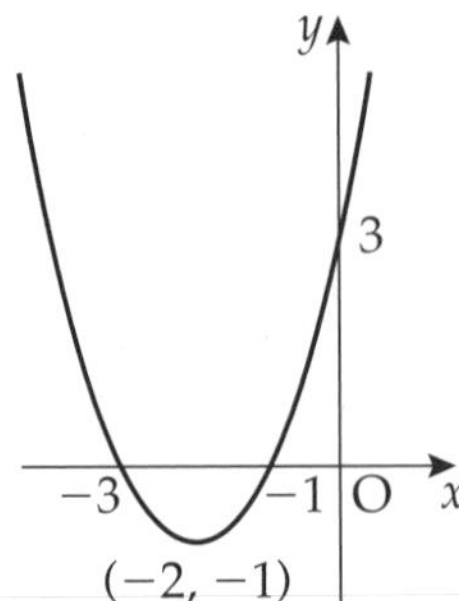

Example continued

Example *continued*

b) Recognise $y = 2^x$ as the equation of an exponential function
$y = 2^x$ has the negative x-axis as an asymptote; it does not cross the x-axis; for crossing the y-axis, $x = 0 \Rightarrow y = 2^0 = 1 \Rightarrow (0, 1)$; also $x = 1 \Rightarrow y = 2^1 = 2 \Rightarrow (1, 2)$

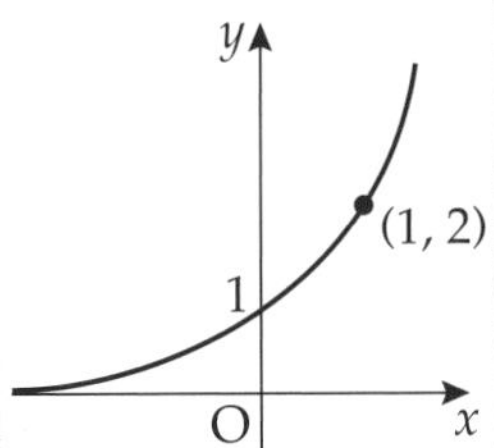

2·7 how to add, subtract, multiply and divide algebraic fractions

Example

Find a) $\frac{a}{x} + \frac{b}{y}$ b) $\frac{x+1}{x-1} - \frac{x-2}{x+2}$ c) $\frac{x^2-x-2}{x^2-x-6} \div \frac{x^2-4x+4}{x^2-4}$.

(Solution) a) both denominators must be the same to add or subtract fractions so multiply top and bottom of the first fraction by y and the second by x

i.e. $\frac{a}{x} + \frac{b}{y} = \frac{ay}{xy} + \frac{bx}{yx} = \frac{ay+bx}{xy}$

b)
$$\frac{x+1}{x-1} - \frac{x-2}{x+2} = \left(\frac{x+1}{x-1} \times \frac{x+2}{x+2}\right) - \left(\frac{x-2}{x+2} \times \frac{x-1}{x-1}\right)$$
$$= \frac{(x+1)(x+2) - (x-2)(x-1)}{(x-1)(x+2)}$$
$$= \frac{(x^2+3x+2) - (x^2-3x+2)}{x^2+x-2} = \frac{6x}{x^2+x-2}$$

c)
- to divide by a fraction, multiply by its multiplicative inverse, e.g. to divide by $\frac{3}{2}$ multiply by $\frac{2}{3}$
- factorise each quadratic expression
- cancel a factor of the top line with the same factor of the bottom line

$$\frac{x^2-x-2}{x^2-x-6} \div \frac{x^2-4x+4}{x^2-4} = \frac{x^2-x-2}{x^2-x-6} \times \frac{x^2-4}{x^2-4x+4}$$
$$= \frac{(x+1)\cancel{(x-2)}}{\cancel{(x+2)}(x-3)} \times \frac{\cancel{(x-2)}\cancel{(x+2)}}{\cancel{(x-2)}\cancel{(x-2)}}$$
$$= \frac{(x+1)}{(x-3)}$$

Chapter 3

THE STRAIGHT LINE

What You Should Know

3·1 that the distance between (x_1, y_1) and (x_2, y_2) is $\sqrt{(x_1 - x_2)^2 + (y_1 - y_2)^2}$ and hence how to calculate the distance between two points in the (x, y) plane.

3·2 how to calculate the co-ordinates of the mid point of a line.

3·3 that the gradient of a line is the tangent of the angle between the line and the positive direction of the x-axis, and how to find the gradient from the angle and vice versa.

3·4 that the gradient of the line joining the points (x_1, y_1) and (x_2, y_2) is $\frac{y_2 - y_1}{x_2 - x_1}$ $\left(\text{or } \frac{y_1 - y_2}{x_1 - x_2}\right)$ and how to calculate the gradient of a line

3·5 that parallel lines have equal gradients

3·6 that lines with non-zero gradients m_1, m_2, are perpendicular $\Leftrightarrow m_1 \times m_2 = -1$, and how to find the gradient of a line perpendicular to a given line, or to determine whether or not two lines are perpendicular.

3·7 how to determine whether or not a set of points is collinear

3·8 that the line with equation $y = mx + c$ passes through $(0, c)$ with gradient m, and how to determine, given the equation of a line, its gradient and its intercept on the y-axis

3·9 that the line through the point (x_1, y_1) with gradient m has equation $y - y_1 = m(x - x_1)$, and how to determine the equation of any line, given its gradient and the co-ordinates of a point on it (or any equivalent information)

3·10 that every straight line has an equation of the form $ax + by + c = 0$ and conversely, every locus of points satisfying an equation of this form is a straight line

3·11 that the point of intersection of two lines is the simultaneous solution of their equations, and hence how to determine the coordinates of the point of intersection of two lines

3·12 how to sketch on a Cartesian diagram (a co-ordinate diagram) the lines with equations of the form

$$x = h,\ y = k,\ y = mx,\ y = mx + c,\ ax + by + c = 0$$

3·13 how to determine whether or not a given set of lines is concurrent and know the standard concurrency results for a triangle

3·14 how to obtain the equation of a median of a triangle (a line from a vertex to the mid point of the opposite side)

3·15 how to obtain the equation of an altitude of a triangle (a line from a vertex perpendicular to the opposite side)

3·16 how to obtain the equation of the perpendicular bisector of a line (a line passing through the mid point and at right angles to the given line)

3·17 how to combine the above to solve more complex problems

Notes on items 3·1 to 3·17: The Straight Line

3·1 that the distance between (x_1, y_1) and (x_2, y_2) is $\sqrt{(x_1 - x_2)^2 + (y_1 - y_2)^2}$ and how to calculate the distance between two points in the (x, y) plane.

- subtract the x-coordinates, subtract the y-coordinates; take care with any negatives
- add the squares of these differences, remember it's just Pythagoras for a hypotenuse
- remember to take the square root; a surd in its simplest form may be required

e.g. A $(3, -4)$ B $(-5, 6)$ gives AB $= \sqrt{(3-(-5))^2 + (-4-6)^2}$
$= \sqrt{8^2 + (-10)^2} = \sqrt{164}$
$= 2\sqrt{41}$

(It doesn't matter whether we subtract $(x_1 - x_2)$ or $(x_2 - x_1)$ etc, since we are squaring the term.)

3·2 how to calculate the co-ordinates of the mid point of a line

Use $\left(\frac{x_1 + x_2}{2}, \frac{y_1 + y_2}{2}\right)$ i.e. (average of the x–co-ordinates, average of the y–co-ordinates)

- remember to ADD the co-ordinates and HALVE the sum
- remember to add two x–co-ordinates and two y–co-ordinates (not one of each)
- this has normally been a one mark question. You are allowed to do it mentally, but do not be careless

e.g. A $(3, -4)$, B $(-5, 6)$ gives a mid point $\left(\frac{3+(-5)}{2}, \frac{-4+6}{2}\right) = (-1, 1)$

3·3 that the gradient of a line is the tangent of the angle between the line and the positive direction of the x-axis, and how to find the gradient from the angle and vice versa

- a line making an angle of 35° with the x-axis has a gradient of tan 35° or 0·7.
- a line with a gradient of 2 makes an angle of $\tan^{-1} 2$ or 63·4° with the x-axis.
- a line with a negative gradient will make an obtuse angle with the (positive) x-axis

e.g. a line at 135° to the x-axis has a gradient of tan 135° or –1.

3·4 that the gradient of the line joining the points (x_1, y_1) and (x_2, y_2) is $\frac{y_2 - y_1}{x_2 - x_1}$ $\left(\text{or } \frac{y_1 - y_2}{x_1 - x_2}\right)$ and hence how to calculate the gradient of a line

e.g. for A $(3, -4)$ B $(-5, 6)$, $m_{AB} = \frac{6-(-4)}{-5-3} = \frac{10}{-8} = -\frac{5}{4}$

- the y-values must be on the top line
- the order on the bottom line must match the order on the top (i.e. B's co-ordinates first both times)
- never just call the gradient m, always indicate the line whose gradient you are calculating, call it m_{AB} as above, for example. (communicate !)

3·5 **that parallel lines have equal gradients.**

(this becomes relevant in **3·7**)

3·6 **that lines with non-zero gradients m_1, m_2, are perpendicular $\Leftrightarrow$ $m_1 \times m_2 = -1$, and how to find the gradient of a line perpendicular to a given line or to determine whether or not two lines are perpendicular.**

! **Given a line with gradient $\frac{p}{q}$, to find the gradient of a perpendicular line, invert $\frac{p}{q}$ and change the sign, i.e. $-\frac{q}{p}$ e.g. 3 and $-\frac{1}{3}$ are gradients of perpendicular lines**

3·7 **how to determine whether or not a set of points is collinear.**

Example

Show that A (2, 5), B (6, 8), and C (14, 14) are collinear.

(Solution) $m_{AB} = \frac{8-5}{6-2} = \frac{3}{4}$ $\quad$ $m_{BC} = \frac{14-8}{14-6} = \frac{3}{4}$

i.e. $m_{AB} = m_{BC}$ $\quad$ $\Rightarrow$ AB $\parallel$ BC

since the point B is common to both lines, A, B, and C are collinear

*do not omit the reference to the common point or you will lose a mark

3·8 **that the line with equation $y = mx + c$ passes through $(0, c)$ with gradient m, and how to determine, given the equation of a line, its gradient and its intercept on the y-axis**

Example

Find the equation of the line passing through (0, 5) with gradient $\frac{2}{3}$.

(Solution) $c = 5$ and $m = \frac{2}{3} \Rightarrow y = \frac{2}{3}x + 5 \Rightarrow 3y = 2x + 15$

Remember: multiply all three terms by 3.

Avoid the common error of writing $3y = 2x + 5$.

Example

Find the gradient of the line with equation $3x + 4y = 5$.

(Solution) Carefully re-arrange the equation $3x + 4y = 5$ into the form $1y = \dots$
(i.e. make y the subject of the formula)

$$4y = -3x + 5 \Rightarrow y = -\frac{3}{4}x + \frac{5}{4} \Rightarrow \text{gradient} = -\frac{3}{4}$$

3·9 that the line through the point (x_1, y_1) with gradient m has equation $y - y_1 = m(x - x_1)$, and how to determine the equation of any line, given its gradient and the co-ordinates of a point on it (or any equivalent information)

(The most common equivalent of a point on the line and its gradient is the co-ordinates of two points. In this case, the gradient can easily be found using the gradient formula, and either point on the line gives the same equation when substituted into the above formula. Other common equivalent data might be a point and the equation of a parallel or perpendicular line, or a point and the intersection of two other lines, etc.)

Example

A is the point (4, 11), B (3, 5) and C (7, 8). Find the equation of the line through A parallel to BC.

(Solution) (Find the gradient of BC, and use it in the formula with the co-ordinates of A. It is essential to use the co-ordinates of a point which lies on the line whose equation you are finding. B and C do not lie on this line.)

$$m_{BC} = \frac{8-5}{7-3} = \frac{3}{4},\ \text{A}\ (4, 11) \Rightarrow y - 11 = \frac{3}{4}(x - 4)$$

It is good practice to simplify the equation of a line, to at most three terms, even though there are various acceptable forms of the answer.

At this point, always clear the fractions before you expand the brackets.

Multiply through by 4 $\Rightarrow 4y - 44 = 3(x - 4) = 3x - 12$

$\Rightarrow 4y = 3x + 32$

3·10 that every straight line has an equation of the form $ax + by + c = 0$ and conversely, every locus of points satisfying an equation of this form is a straight line

An equation does **not** represent a straight line if it includes other powers of x or y, e.g. $x^2, xy, y^3, \frac{1}{x}$. A locus of points is a set of points, usually with a common property.

3·11 **that the point of intersection of two lines is the simultaneous solution of their equations, and hence how to determine the co-ordinates of the point of intersection of two lines.**

Item **2·1** shows that the lines with equations $2x + 5y + 14 = 0$ and $3x - y = 13$ intersect at the point $(3, -4)$.

3·12 **how to sketch on a Cartesian diagram (a co-ordinate diagram) the lines with equations of the form $x = h$, $y = k$, $y = mx$, $y = mx + c$, $ax + by + c = 0$**

Example

Sketch the lines with equations

a) $x = 2$ b) $y = 3$ c) $y = 4x$ d) $y = 4x + 5$ e) $3x + 4y = 12$.

(Solution) **a)** and **b)** are easy but they can still cause confusion in exams!

c) and **d)** have gradient 4
c) passes through the origin
d) passes through $(0, 5)$

Remember that
$x = h$ is parallel to the y-axis
$y = k$ is parallel to the x-axis

Remember that
$y = mx$ passes through O with gradient m
$y = mx + c$ has gradient m and intercept c

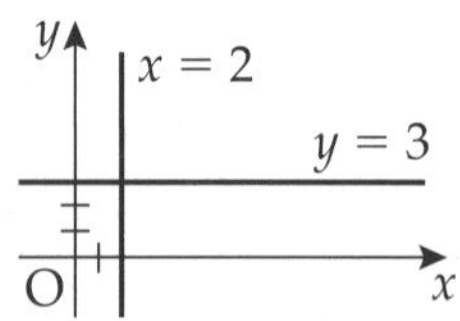

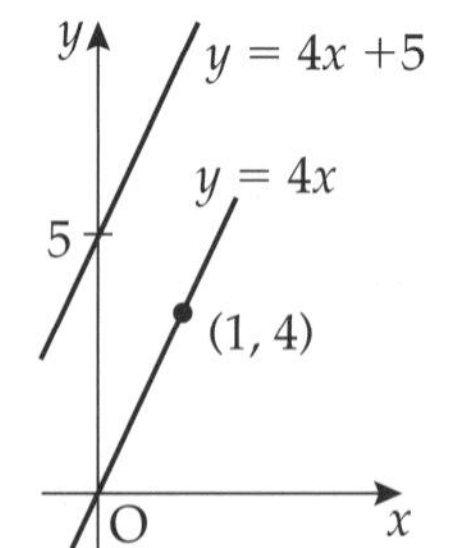

e) the easiest way to sketch this line is to find where it crosses the axes:
for crossing Oy, $x = 0 \Rightarrow y = 3$
hence $(0, 3)$
for crossing Ox, $y = 0 \Rightarrow x = 4$
hence $(4, 0)$

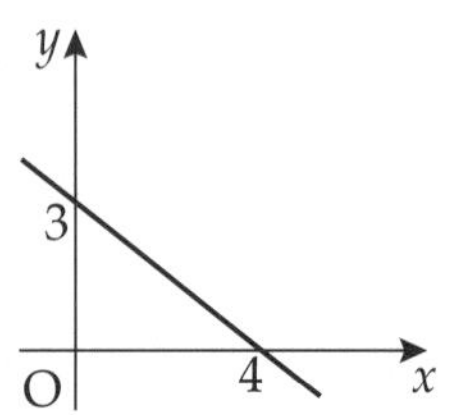

3·13 how to determine whether or not a given set of lines is concurrent and know the standard concurrency results for a triangle

Example

Show that the lines with equations $2x + 5y + 14 = 0$, $3x - y = 13$ and $7x + 5y = 1$ are concurrent.

(Solution) **We do not need to solve each possible pair of equations. Simply pick the two easiest looking equations, find where these lines meet and see if that point lies on the third line, by checking that its co-ordinates satisfy the equation.**

We saw in item 3·11 that the first two lines intersect at the point $(3, -4)$.
Substitute in the third equation: $7x + 5y = 7(3) + 5(-4) = 21 - 20 = 1$
i.e. the point of intersection of the first two lines lies on the third line, so the three lines are concurrent, at the point $(3, -4)$.

Standard concurrency results for a triangle:

Key Words and *Definitions*

1 the medians are concurrent at the *centroid*

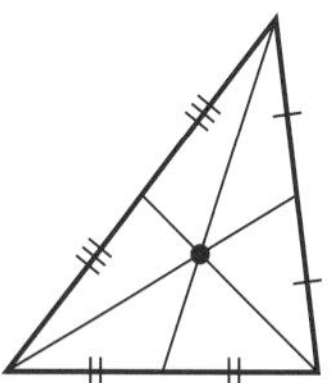

2 the altitudes are concurrent at the *orthocentre*

3 the perpendicular bisectors of the sides are concurrent at the *circumcentre*
the *circumcircle* passes through the vertices

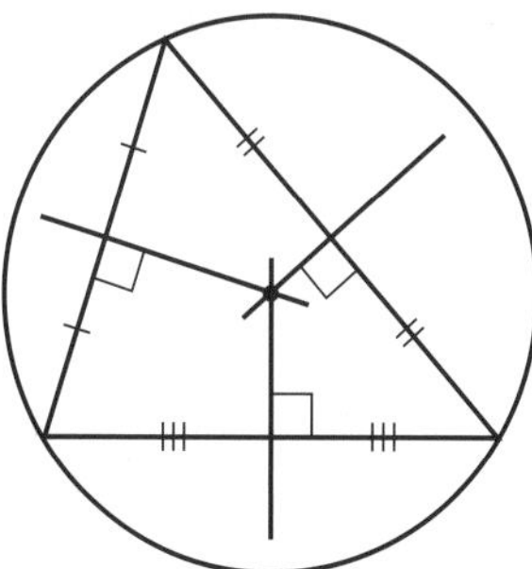

4 the bisectors of the angles are concurrent at the *in-centre*
the sides are tangents to the *in-circle*

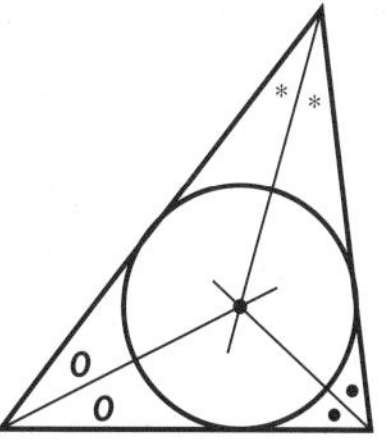

3·14 how to obtain the equation of a median of a triangle (a line from a vertex to the mid point of the opposite side)

Example

Triangle PQR has vertices P $(-1, 3)$, Q $(4, 2)$ and R $(-2, -4)$. Find the equation of the median PM.

(Solution) You have been given the co-ordinates. Make a sketch on plain paper. Make it reasonably accurate without it being a scale drawing. Then you can check if you make any glaring errors, especially with the signs of gradients

M is $\left(\frac{4+(-2)}{2}, \frac{2+(-4)}{2}\right) = (1, -1)$

M $(1, -1)$, P $(-1, 3)$

$\Rightarrow m_{PM} = \frac{3-(-1)}{-1-1} = \frac{4}{-2} = -2$

using $y - y_1 = m(x - x_1)$

$\Rightarrow$ PM has equation $y - (-1) = -2\,(x - 1)$
which simplifies to $2x + y = 1$

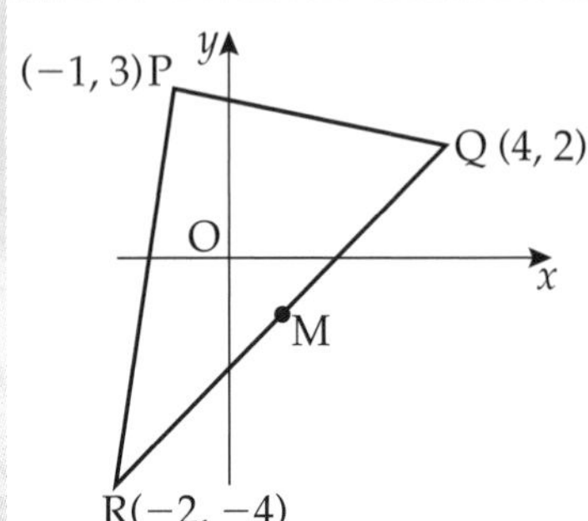

Note: 1 to find the equation of PM, we could have used the co-ordinates of either P or M in $y - y_1 = m(x - x_1)$ since we knew both

2 PM and QR are not perpendicular (confusion between medians, altitudes and perpendicular bisectors is common.)

3·15 how to obtain the equation of an altitude of a triangle (a line from a vertex perpendicular to the opposite side)

Example

In triangle PQR defined in 3·14, find the equation of the altitude PN.

(Solution) $m_{QR} = \frac{2-(-4)}{4-(-2)} = \frac{6}{6} = 1$

$\Rightarrow m_{PN} = -1$ (using $m_1 \times m_2 = -1$)

using P $(-1, 3)$ in $y - y_1 = m(x - x_1)$
gives $y - 3 = -1\,(x - (-1))$
$\Rightarrow x + y = 2$

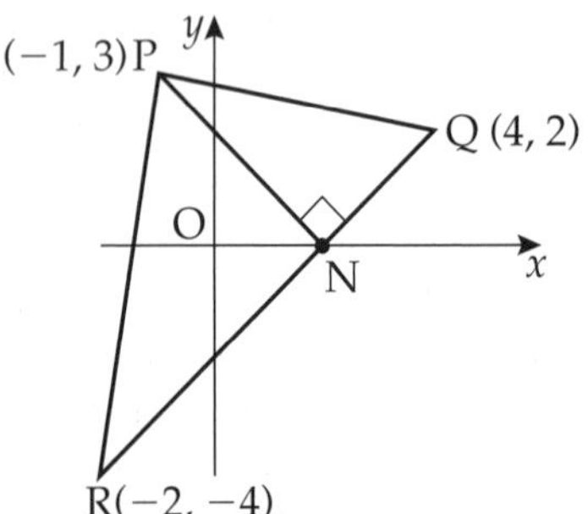

Note: 1 we had to use the co-ordinates of P since we did not know those of N

2 N is not the mid point of RQ (do not confuse the altitude with the median)

3·16 how to obtain the equation of the perpendicular bisector of a line (a line passing through the mid point and at right angles to the given line)

Example

Find the perpendicular bisector of KL where K is (4, 5) and L is (8, −3)

(Solution) For this we need the co-ordinates of M, the mid point of KL and the gradient of a line perpendicular to KL.

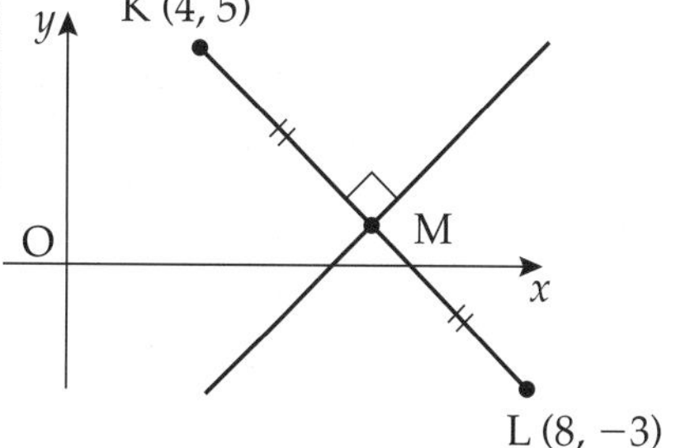

M is $\left(\frac{4+8}{2}, \frac{5+(-3)}{2}\right) = (6, 1)$,

$m_{KL} = \frac{5-(-3)}{4-8} = \frac{8}{-4} = -2$

$\Rightarrow m_{\perp} = \frac{1}{2}$ (using $m_1 \times m_2 = -1$)

$\therefore\ y - 1 = \frac{1}{2}(x - 6) \Rightarrow 2y - 2 = x - 6 \Rightarrow 2y = x - 4$

3·17 how to combine the above to solve more complex problems

(See Chapter 18 – Typical longer questions: Question 18·1.)

Arguably more marks are lost through making casual errors than not knowing how to do a question. So error avoidance is the name of the game.

Common Mistakes

The Straight Line is a straightforward topic but pupils still lose marks through errors. What errors should be avoided?

Copying co-ordinates or equations down wrongly.
Getting the signs wrong when using the distance formula.
Getting the gradient upside down.
Making errors in subtracting negative numbers.
Subtracting instead of adding when getting a mid point.
Forgetting how to obtain the equations of lines parallel to the axes.
Confusing medians, altitudes and perpendicular bisectors.
Not knowing the meaning of the word concurrent.
Not including 'common point, common direction' in a collinearity proof.
Not re-arranging an equation correctly.
Not drawing your own sketch, which would check many calculations.
Using the co-ordinates of a point which is not on the line whose equation you are finding.

Check each line before you move on to the next one. Full marks ahead.

Chapter 4

FUNCTIONS AND GRAPHS

What You Should Know

4·1 the meaning of the terms domain, codomain, range, image, function, composite function, inverse function, and the notation f^{-1}

4·2 when an arrow diagram or a graph represents a relation, a function, or a 1–1 correspondence

4·3 how to determine the range of a given function

4·4 how to obtain a formula for the composition of two given functions (i.e. find $f(g(x))$ given the formulae for $f(x)$ and $g(x)$)

4·5 the conditions for a function to have an inverse

4·6 that the graphs of a function and its inverse are images under reflection in the line $y = x$

4·7 the general features of the graphs of $y = a^x$ and $y = \log_a x$

4·8 given the graph of a function f, how to sketch the graphs of related functions e.g. $y = -f(x)$, $y = 3f(x)$, $y = f(x)+2$, $y = f(x-1)$, $y = f'(x)$

4·9 how to complete the square on a quadratic expression

4·10 how to interpret formulae and equations

4·11 how to recognise the probable form of a function from its graph

Notes on items 4·1 to 4·11: Functions and Graphs

4·1 the meaning of the terms domain, codomain, range, image, function, composite function, inverse function, and the notation f^{-1}

Key Words and Definitions

An arrow diagram is a useful way to show a simple function e.g.

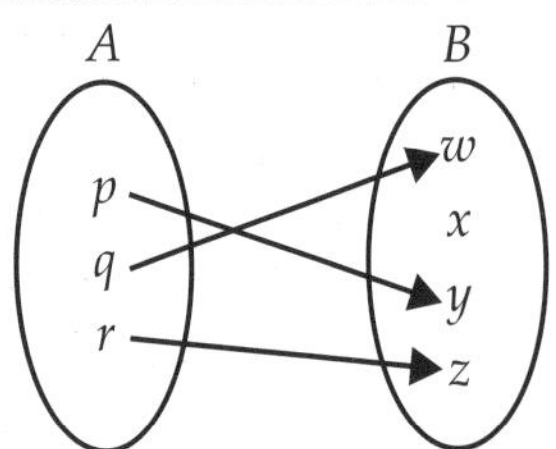

$A = \{p, q, r\}$ is the **domain**
$B = \{w, x, y, z\}$ is the **codomain**
w is the **image** of q, for example
$\{w, y, z\}$ is the **range** (the set of images)

This is a **function** because each element of A has exactly one image in B.
See 4·4 for **composite function** and 4·2 for **inverse function** and the notation f^{-1}.

4·2 when an arrow diagram or a graph represents a relation, a function, or a 1–1 correspondence

Key Points

arrow diagrams:

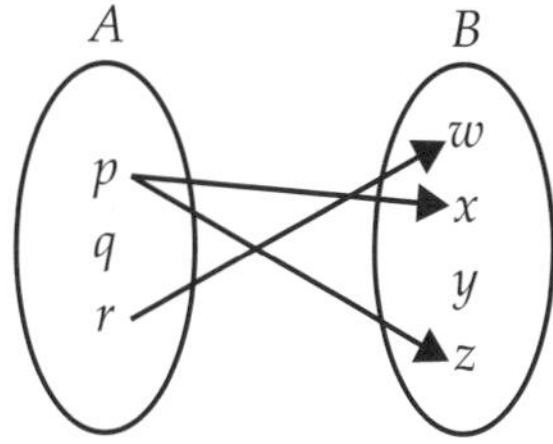

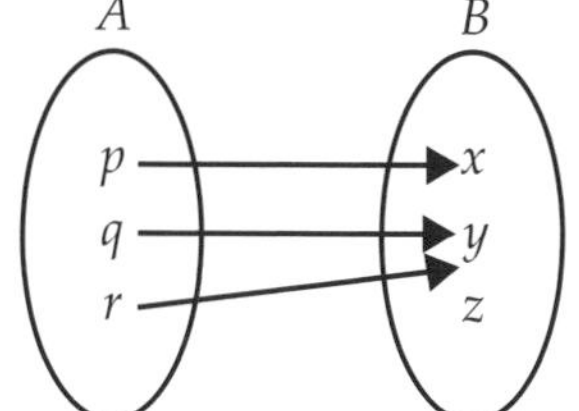

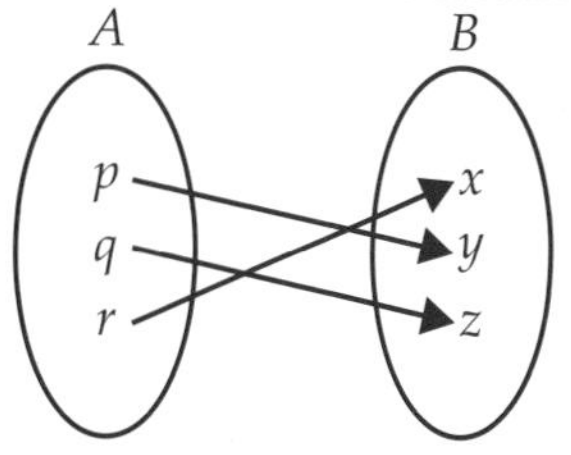

This is only a **relation**. It is not a function for either of two reasons: q has no image, or p has two images.

This is a **function**. Each element of A has exactly one image in B. It is not a 1–1 correspondence for either of two reasons: q and r both map to y or z is 'left over'.

This is a **1-1 correspondence**. All the elements of A and B are paired off . It is a function where separate elements of A map to separate elements of B and the range is the whole of the codomain

If, in a 1–1 correspondence, the arrows are reversed, another 1–1 correspondence is obtained. This is called the **inverse function** and is denoted by f^{-1}.

graphs:

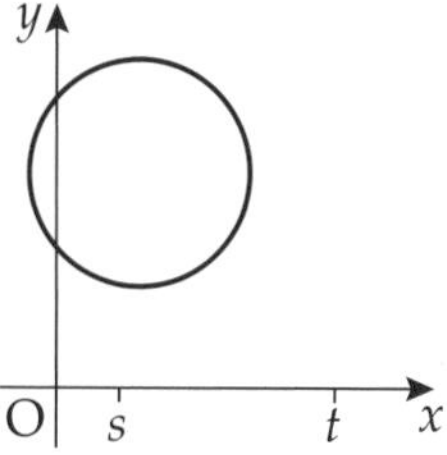

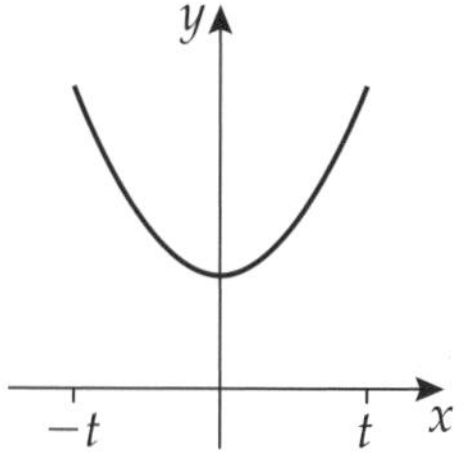

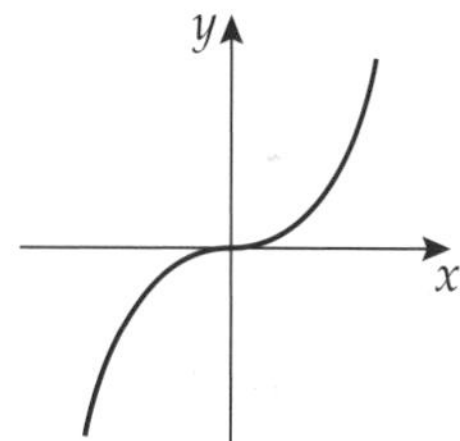

This is only a **relation**. It is not a function for either of two reasons: t, for example, has no image, or s has two images.

This is a **function**. Each element of the domain (i.e. the x-axis) has exactly. one image in the codomain (i.e. the y-axis) It is not a 1–1 correspondence for either of two reasons: t and $(-t)$ both map to the same y-value or none of the negative values in the codomain are images of anything in the domain.

This is a **1–1 correspondence**. All the elements of the domain (i.e. the x-axis) and the codomain (i.e. the y-axis) are paired off. It is a function with two properties: Separate elements of the domain map to separate elements of the codomain and the range is the whole of the codomain.

4·3 how to determine the range of a given function

Example

Find the range of the functions

a) $f(x) = x^2 + 1$ b) $g(x) = \sqrt{1 - x^2}$ c) $h(x) = \sin(x°)$ d) $l(x) = 2x$

(Solution)

a) use the graph the whole graph is above the line $y = 1$ so the range is $y \geqslant 1$

b) only the square root of a positive quantity exists, so the domain is is restricted to $-1 \leqslant x \leqslant 1$ $x = \pm 1 \Rightarrow y = 0$ $x = 0 \Rightarrow y = 1$, so the range is $0 \leqslant y \leqslant 1$

c) recall the sine graph; it has a maximum of 1 and a minimum of -1. so the range is $-1 \leqslant y \leqslant 1$

d) the graph of $y = 2x$ goes all the way up and down the page so the range is all real numbers.

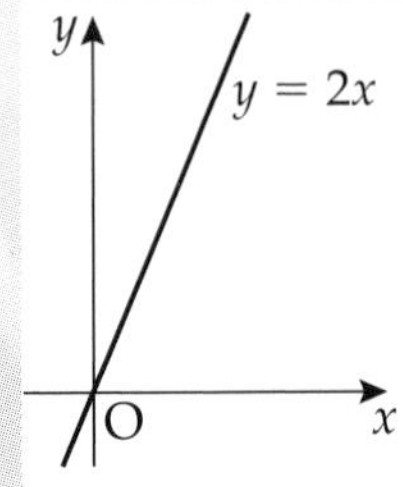

4·4 how to obtain a formula for the composition of two given functions (i.e. find $f(g(x))$ given the formulae for $f(x)$ and $g(x)$)

In examples like these, the domain and codomain of both functions are usually the set of all real numbers. To avoid any problem arising from restrictions on the given functions, exam questions often say that the functions are 'defined on suitable domains'.

Example

Three functions f, g and h are defined on suitable domains by

$f(x) = 2x + 1$, $g(x) = 3x - 2$ and $h(x) = \frac{1}{2}(x - 1)$

a) Find the formula for $f(g(x))$.

b) Show that f and h are inverses of each other.

(Solution) a) $f(g(x)) = f(3x - 2)$ include this line – it could be worth a mark

$= 2(3x - 2) + 1$ think: '*f*(*something*) = 2(*something*) + 1'

$= 6x - 3$

b) $f(h(x)) = f(\frac{1}{2}(x - 1))$

$= 2(\frac{1}{2}(x - 1)) + 1$

$= x - 1 + 1$

$= x = i(x)$ where $i(x)$ = the identity function; hence f and h are inverses of each other. [e.g. $h(5) = 2$ and $f(2) = 5$]

4·5 the conditions for a function to have an inverse

Only a one to one correspondence can have an inverse. [Sometimes we have to be careful in how we define the domain and codomain to ensure that we have a one to one correspondence, e.g. in item 4·7.]

Remember

In an arrow diagram, the elements of the domain and codomain must be paired off. Reversing the direction of the arrows gives the inverse function.

Instead of 'x maps to y under f' we get 'y maps to x under f^{-1}', e.g.

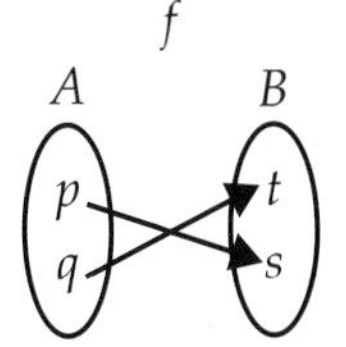

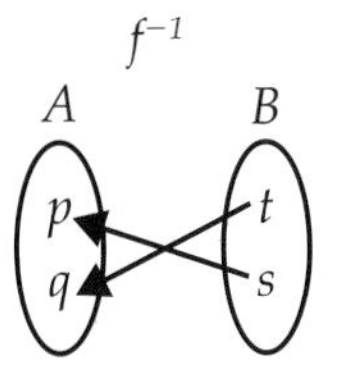

On a graph, the graph must go all the way across the page (for it to be a function), and it must go all the way up and down the page (for the range to be the whole of the codomain) but it must never be at the same height more than once (for it to be 1–1).

If (x, y) is on the graph of $y = f(x)$ then (y, x) is on the graph of the inverse, e.g.

alternatively the point $(x, f(x))$ is on the graph of $y = f(x)$ and the point $(f(x), x)$ is on the graph of the inverse function.

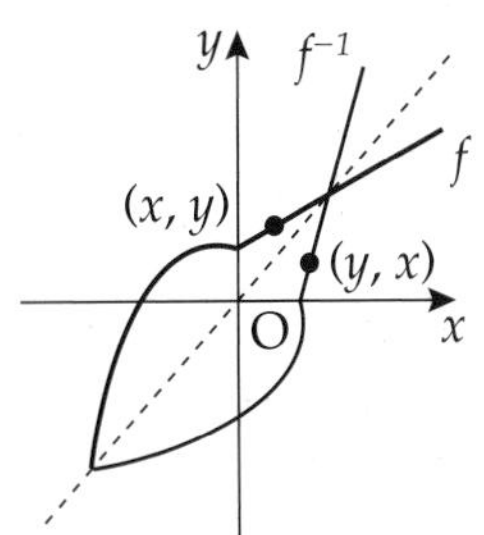

Given the formula for a function, finding a formula for the inverse function is not part of the syllabus, but is still a useful exercise to understand before you meet logarithms. You can still, however, be asked questions like 4·4b.

4·6 that the graphs of a function and its inverse are images under reflection in the line $y = x$

In the diagram for 4·5, the graphs of f and f^{-1} are reflections of each other in the line $y = x$.

Example

Show that the functions f and g are inverses of each other where $f(x) = 2x + 2$ and $g(x) = \frac{1}{2}(x - 2)$ for x any real number, and sketch the graphs of both on the same cartesian diagram.

(Solution)

$$\begin{aligned} f(g(x)) &= f\left(\frac{1}{2}(x-2)\right) \\ &= 2\left(\frac{1}{2}(x-2)\right) + 2 \\ &= x - 2 + 2 \\ &= x \end{aligned}$$

$\Rightarrow f$ and g are inverses of each other

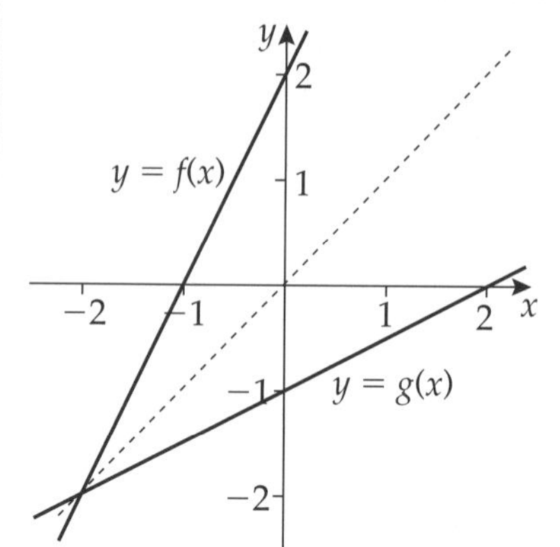

4·7 the general features of the graphs of $y = a^x$ and $y = \log_a x$

Key Points

$y = a^x \ (a > 1)$

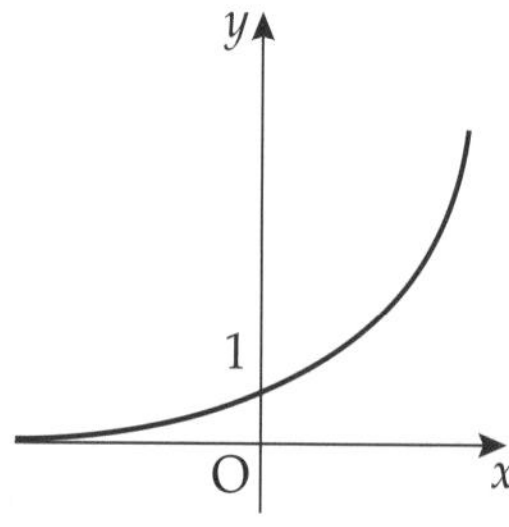

The negative x-axis is the asymptote.
The function is always increasing.
For all values of a, the graph passes through the point $(0, 1)$
The exponential graph is really revision. Look back at item **2·6** for $y = 2^x$.

$y = \log_a x \ (a > 1)$

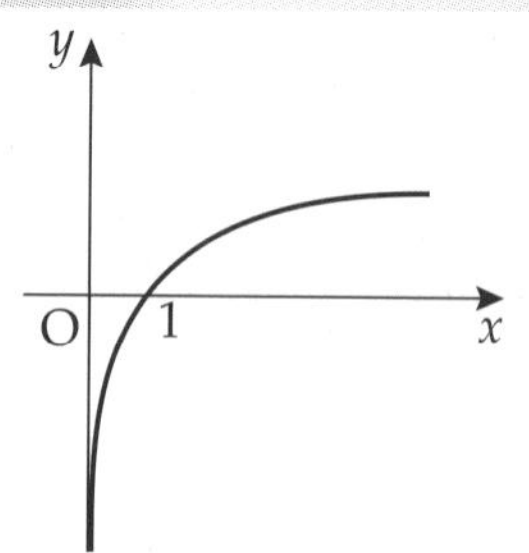

The negative y-axis is the asymptote.
The function is always increasing.
For all values of a, the graph passes through the point $(1, 0)$
The function is only defined for $x > 0$, i.e. the domain is $\{x \in \mathbb{R}, x > 0\}$

We create the logarithm function as the inverse of the exponential function. To ensure that exp(x) is a 1–1 correspondence, we carefully define the codomain as the set of all non-negative real numbers. Thus we have exp: $x \rightarrow a^x$
and $\log_a : a^x \rightarrow x$.

***Key Points** continued* ➢

Key Points continued

We cannot expect all real numbers to be written in the form 'a' to the power 'something', so we also write the definition of a logarithm as follows

$N = b^x \Leftrightarrow \log_b N = x$

As these are inverse functions of each other, their graphs are images of each other under reflection in the line with equation $y = x$.

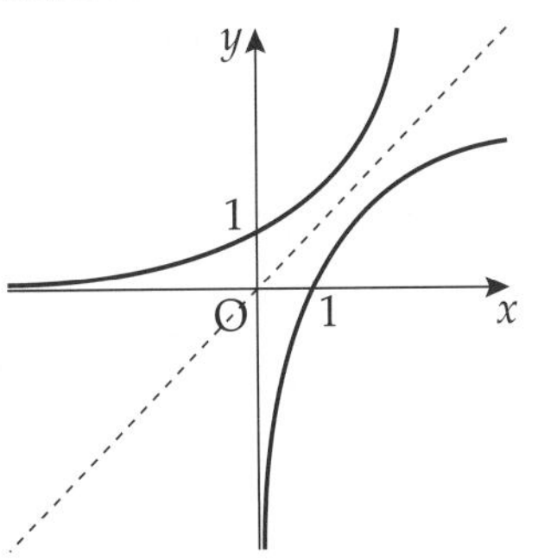

4·8 given the graph of a function f, how to sketch the graphs of related functions e.g. $y = -f(x)$, $y = 3f(x)$, $y = f(x) + 2$, $y = f(x-1)$, $y = f'(x)$

Key Points

$y = -f(x)$	reflect $y = f(x)$ in the x-axis
$y = f(-x)$	reflect $y = f(x)$ in the y-axis
$y = -f(-x)$	give $y = f(x)$ a half turn rotation about the origin O (or reflection in the x-axis followed by reflection in the y-axis) [Do not call this reflection in O. Reflection in a point does not exist]
$y = f(x) + a$	slide $y = f(x)$ a units up (parallel to the y-axis)
$y = f(x - a)$	slide $y = f(x)$ a units to the right (parallel to the x-axis)
$y = a\,f(x)$	stretch $y = f(x)$ parallel to the y-axis by a scale factor of a
$y = f\left(\frac{x}{a}\right)$	stretch $y = f(x)$ parallel to the x-axis by a scale factor of a
$y = f^{-1}(x)$	reflect in $y = x$

(You may be asked to combine two or more of these transformations. If you translate an exponential or logarithmic function, remember to indicate the new position of the asymptote.)

The final related function (below) could equally well have been dealt with under the topic of differentiation, namely, the graph of the derived function.

$y = f'(x)$: a stationary point at (a, b) on f means $(a, 0)$ is on f';
where f is increasing, the graph of f' is positive, i.e. above the x-axis;
the degree of f' is one less than the degree of the polynomial f.

[The easy cases to remember are that the derivative of a quadratic function (whose graph is a parabola) is a linear function (a straight line graph), and the graph of the derivative of a cubic function (with a max and a min turning point) is a parabola.]

Example

These diagrams show the graphs of

a) $y = f(x) = x^2 - 4x$ b) $y = e^x$ c) $y = \log_5 x$

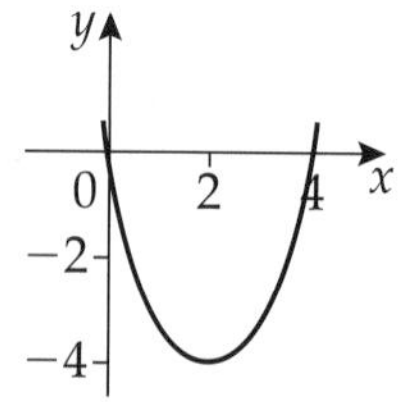

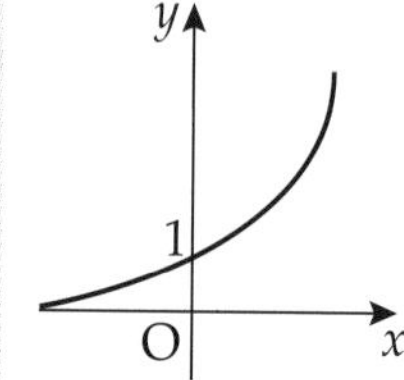

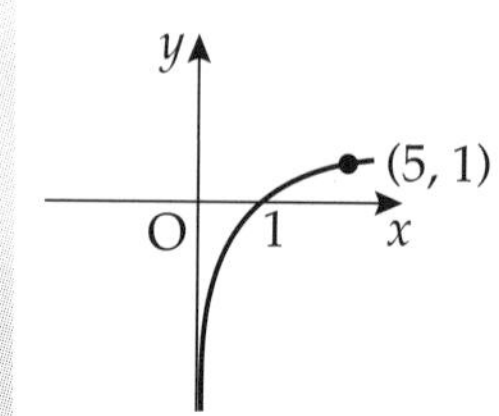

Sketch the graphs of

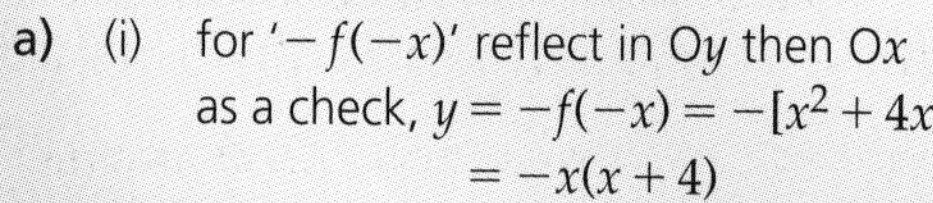

a) (i) $y = g(x) = -f(-x)$ b) $y = 1 + e^{-x}$ c) $y = 1 - \log_5(x - 2)$.
(ii) $y = f'(x)$

(Solution)

a) (i) for '$-f(-x)$' reflect in Oy then Ox
as a check, $y = -f(-x) = -[x^2 + 4x]$
$= -x(x + 4)$
which has zeros at 0 and -4 and has a maximum turning point

(ii) minimum turning point at $(2, -4)$ means f' crosses the x-axis at $(2, 0)$.
for $x < 2$, f decreasing means $f' < 0$
f is quadratic means f' is linear
[I based this working solely on the graph of f, not on the fact that $f'(x) = 2x - 4$.]

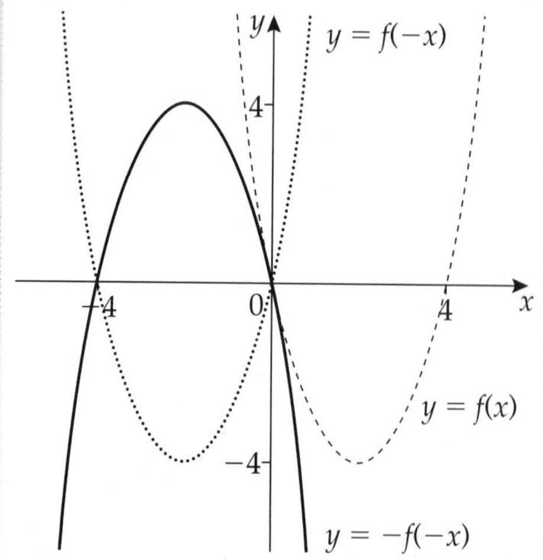

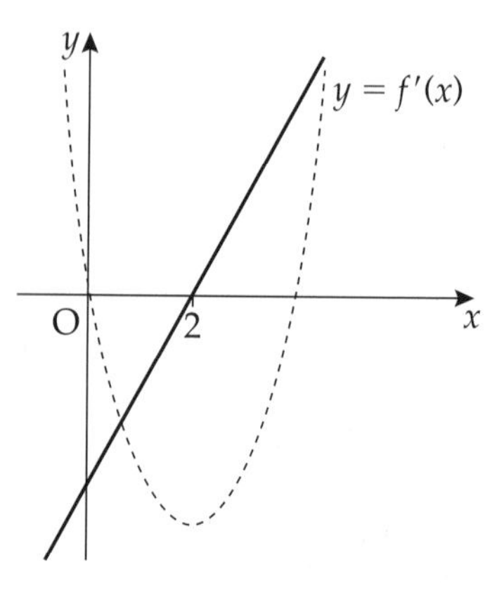

Example continued ➢

Example continued

b) $y = e^{-x}$ is $y = e^x$ reflected in Oy;
$y = 1 + e^{-x}$ is $y = e^{-x}$ translated 1 up

c) reflect in Ox to get $-\log_5 x$; translate 2 units right to get $-\log_5(x - 2)$; translate 1 unit up to get $1 - \log_5(x - 2)$

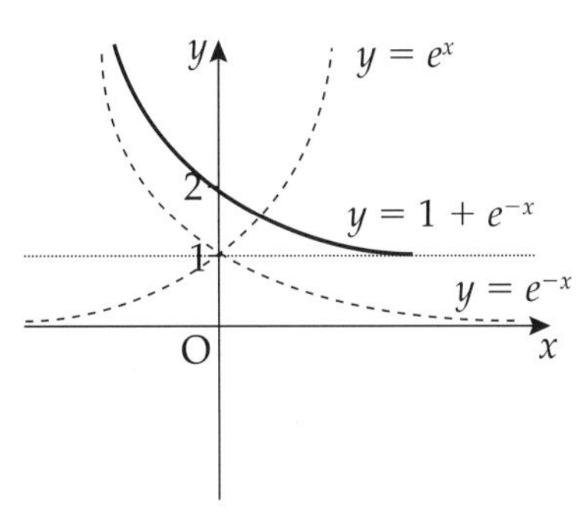

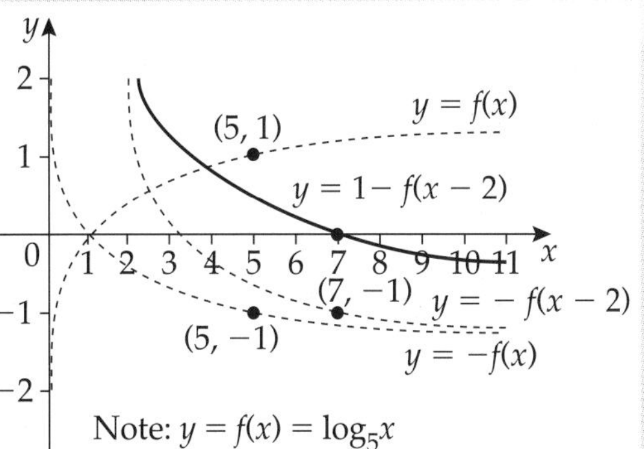

4·9 how to complete the square on a quadratic expression

Any quadratic expression can be written as the sum (or difference) of a number and the square of a binomial (a bracket with two terms), e.g. $5 + (x - 2)^2$.

We do this by separating the terms involving x, adding the number which makes these into a perfect square, and adjusting the constant to preserve the size of the original expression.

This depends on the algebraic identities $(x + a)^2 = x^2 + 2ax + a^2$, **and** $(x - a)^2 = x^2 - 2ax + a^2$.

These identities show that the crucial aspect is that the constant term, which is a^2, is the square of half of the coefficient of x, which is $\pm 2a$.

Example

Express
a) $x^2 + 10x + 12$ in the form $(x + a)^2 + b$
b) $3 + 4x - x^2$ in the form $c - (x + d)^2$
c) $2x^2 + 4x + 5$ in the form $p(x + q)^2 + r$.

(Solution)

a)
$x^2 + 10x + 12$
$= (x^2 + 10x) + 12$
$= (x^2 + 10x + 25 - 25)^* + 12$
$= (x^2 + 10x + 25) - 25 + 12$
$= (x + 5)^2 - 13$
*25 is (half of 10)^2

b)
$3 + 4x - x^2$
$= 3 - (x^2 - 4x)$
$= 3 - (x^2 - 4x + 4 - 4)$
$= 3 - (x^2 - 4x + 4) + 4$
$= 7 - (x - 2)^2$
watch the $-(-4)$ becoming $+4$

c)
$2x^2 + 4x + 5$
$= 2(x^2 + 2x) + 5$
$= 2(x^2 + 2x + 1 - 1) + 5$
$= 2(x^2 + 2x + 1) - 2 + 5$
$= 2(x + 1)^2 + 3$
always work with $1x^2$

Example continued ➢

Example continued

In practice you would not need to show every line of this working. Having written these quadratic functions in these forms, it is easy to see that

a) has a minimum value of -13 when $x = -5$
b) has a maximum value of 7 when $x = 2$
c) has a minimum value of 3 when $x = -1$.

These facts are useful for sketching the parabola which is the graph of each function, especially where the quadratic does not factorise.

Completing the square is not just an abstract exercise. Its purpose is to identify maximum or minimum turning points, which helps in sketching the graph of the related parabola.

for example $y = 3(x-1)^2 + 2$ has a minimum value of 2 when $x = 1$,
$y = 18 - 2(x-4)^2$ has a maximum value of 18 when $x = 4$

The quadratic written in the form of a completed square can also be used to solve the equation for where the graph crosses the x-axis, but remember the $\pm$ when you take the square root of both sides,
e.g. crosses the x-axis where $y = 0$,

$$\Rightarrow 18 - 2(x-4)^2 = 0 \Rightarrow (x-4)^2 = 9 \Rightarrow (x-4) = \pm 3 \Rightarrow x = 4 \pm 3 = 7{,}1$$

4·10 how to interpret formulae and equations

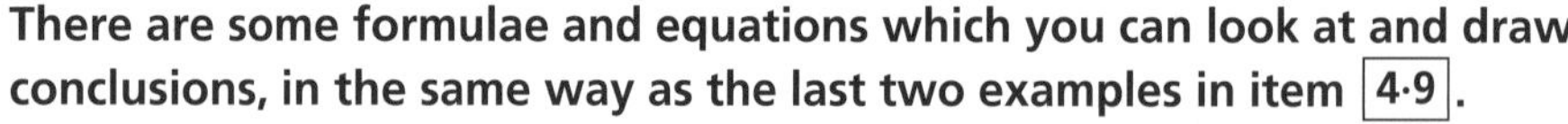

There are some formulae and equations which you can look at and draw conclusions, in the same way as the last two examples in item 4·9.

For example:

$y = \sqrt{9 - x^2}$	**only exists for $-3 < x < 3$ (you cannot take the square root of anything negative);**
$y = \dfrac{2}{x-1}$	**is not defined at $x = 1$; the graph has a discontinuity there (you cannot divide by 0);**
$y = 4\cos(x-30)° + 2$	**has a maximum of 6 when $x = 30$ and a minimum of -2 when $x = 210$**
$y = 1 + 3x^2 - 5x^4$	**possesses symmetry about the y-axis (because all the powers are even) (this is a useful but non-examinable fact)**

4·11 how to recognise the probable form of a function from its graph

Example

Find the equation of each of these graphs.

a)

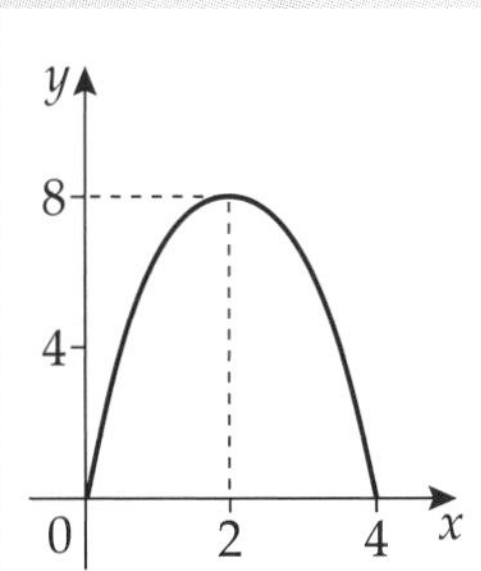

b)

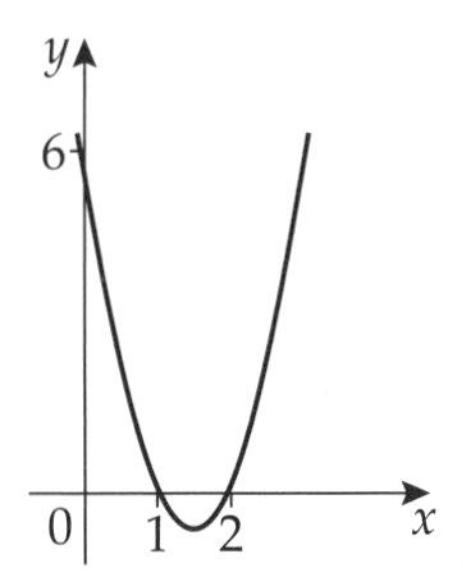

c)

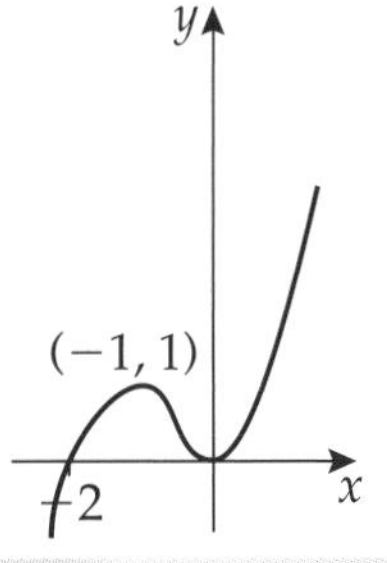

(Solution) a) parabola shape indicates a quadratic function;
roots at $0, 4 \Rightarrow y = kx(x - 4)$ [Do NOT omit the k]
$(2, 8) \Rightarrow 8 = k(2)(-2) \Rightarrow k = -2$
$\Rightarrow y = -2x(x - 4)$
$\Rightarrow y = 8x - 2x^2$

b) parabola shape indicates a quadratic function;
roots at $1, 2 \Rightarrow y = k(x - 1)(x - 2)$
$(0, 6) \Rightarrow 6 = k(-1)(-2) \Rightarrow k = 3$
$\Rightarrow y = 3(x - 1)(x - 2)$
$\Rightarrow y = 3(x^2 - 3x + 2)$
$\Rightarrow y = 3x^2 - 9x + 6$

c) the shape indicates a cubic function;
the x-axis is a tangent at the origin gives a double root at $x = 0$;
roots at $-2, 0, 0 \Rightarrow y = k(x + 2)x^2$
$(-1, 1) \Rightarrow 1 = k(1)(1) \Rightarrow k = 1$
$\Rightarrow y = 1x^2\,(x + 2)$
$\Rightarrow y = x^3 + 2x^2$

Chapter 5

DIFFERENTIATION

What You Should Know

5·1 that the derivative of a function is a measure of the gradient of the graph of the function or the rate of change of the function

5·2 that the derivative is defined by $f'(x) = \lim_{h \to 0} \frac{f(x+h) - f(x)}{h}$

5·3 that the derivative of a) ax^n is nax^{n-1}
b) $f(x) + g(x)$ *is* $f'(x) + g'(x)$
c) $kf(x)$ is $kf'(x)$

5·4 how to apply this rule to differentiate products and quotients such as $(x+1)(x-3)$ or $\frac{x^2 + 5x - 1}{2\sqrt{x}}$

5·5 that the gradient of a curve at a point on it is the gradient of the tangent to the curve at that point

5·6 how to find the equation of the tangent to a curve at a point with given x-co-ordinate

5·7 how to find the points on a curve where the gradient has a given value

5·8 how to determine whether a curve is increasing, decreasing, or stationary at a given point and in an interval

5·9 how to find the stationary points on a curve and determine their nature (i.e. whether maximum or minimum turning point, or point of inflection)

5·10 how to sketch a curve with a given equation and interpret the resulting graph

5·11 how to determine the greatest and least values of $y = f(x)$ on the interval $a \leqslant x \leqslant b$

5·12 how to solve problems requiring optimisation

Notes on items 5·1 to 5·12: Differentiation

5·1 that the derivative of a function is a measure of the gradient of the graph of the function or the rate of change of the function.

It is also known as the derived function and the notations used include $f'(x)$, f', y', $y'(x)$, $\frac{dy}{dx}$, $\frac{df}{dx}$.

$f'(a)$ is the rate of change of f at a or the gradient of the tangent to $y = f(x)$ at $x = a$.
The process of obtaining the derivative is called differentiating.

5·2 **that the derivative is defined by $f'(x) = \lim_{h \to 0} \frac{f(x+h) - f(x)}{h}$**

We rely on this formula later to find the derivatives of $\sin x$ and $\cos x$.

5·3 **that the derivative of**
a) ax^n is nax^{n-1}
b) $f(x) + g(x)$ is $f'(x) + g'(x)$
c) $kf(x)$ is $kf'(x)$

Example

Differentiate a) $8x^{\frac{3}{4}} + 3$ b) $x^{-\frac{2}{3}}$ c) $\frac{1}{2\sqrt{x}}$.

(Solution) a) (multiply the 8 by the three quarters and reduce the three quarters by 1; think 'three quarters minus four quarters gives negative one quarter'; the derivative of any constant is zero (since the gradient of $y = k$ is 0))

so $f(x) = 8x^{\frac{3}{4}} + 3 \Rightarrow f'(x) = 6x^{-\frac{1}{4}}$

b) (the (unwritten) coefficient is 1; multiply by negative two thirds and reduce the two thirds by one, think '−2 thirds − 3 thirds = −5 thirds'

so $f(x) = x^{-\frac{2}{3}} \Rightarrow f'(x) = -\frac{2}{3}x^{-\frac{5}{3}}$

c) **be careful here, for three reasons.**

1 **You cannot differentiate with the x appearing on the bottom line. You need to use the laws of indices**

2 **Deal with the numbers separately; the coefficient is one half.**

3 **Do not confuse converting into an appropriate form with differentiating.**

so $f(x) = \frac{1}{2\sqrt{x}} = \frac{1}{2x^{\frac{1}{2}}} = \frac{1}{2}x^{-\frac{1}{2}} \Rightarrow f'(x) = -\frac{1}{4}x^{-\frac{3}{2}}$

5·4 **how to apply this rule to differentiate products and quotients**

[You can only differentiate sums of powers of x, not products and quotients, so we have to manipulate these into an appropriate form before we can differentiate.]

Example

Differentiate a) $(x + 1)(x - 3)$ b) $\frac{x^2 + 5x - 1}{2\sqrt{x}}$

(Solution) a) You have to expand the brackets first before you can differentiate.

$f(x) = (x + 1)(x - 3) = x^2 - 2x - 3$

$\Rightarrow f'(x) = 2x - 2$

***Example** continued* ➢

Example continued

b) (Each term on the top line must be divided by the bottom line first, but it is usually easier to keep the half as a common factor.)

$$f(x) = \frac{x^2 + 5x - 1}{2\sqrt{x}} = \frac{1}{2}\left[\frac{x^2}{x^{\frac{1}{2}}} + \frac{5x}{x^{\frac{1}{2}}} - \frac{1}{x^{\frac{1}{2}}}\right] = \frac{1}{2}\left[x^{\frac{3}{2}} + 5x^{\frac{1}{2}} - x^{-\frac{1}{2}}\right]$$

Compare this working with that in item **2·2**.

$$\Rightarrow f'(x) = \frac{1}{2}\left[\frac{3}{2}x^{\frac{1}{2}} + \frac{5}{2}x^{-\frac{1}{2}} + \frac{1}{2}x^{-\frac{3}{2}}\right] = \frac{1}{4}\left[3x^{\frac{1}{2}} + 5x^{-\frac{1}{2}} + x^{-\frac{3}{2}}\right]$$

5·5 that the gradient of a curve at a point on it is the gradient of the tangent to the curve at that point

Example

Find the gradient of the curve with equation $y = 4\sqrt{x}$ at the point where $x = 4$.

(Solution) $y = 4\sqrt{x} = 4x^{\frac{1}{2}} \Rightarrow \frac{dy}{dx} = 2x^{-\frac{1}{2}}$

so the gradient at $x = 4$ is $y'(4) = 2(4)^{-\frac{1}{2}} = 2(2^2)^{-\frac{1}{2}} = 2(2)^{-1} = \frac{2}{2} = 1$

5·6 how to find the equation of the tangent to a curve at a point with given x-co-ordinate

Example

Find the equation of the tangent to the curve with equation $y = x^2 - 4x + 5$ at the point where it crosses the y-axis.

(Solution) the curve crosses the y-axis where $x = 0$, i.e. at the point $(0, 5)$;
$y = x^2 - 4x + 5 \Rightarrow y'(x) = 2x - 4$;
the gradient at $(0, 5) = y'(0) = -4$;
hence the equation of the tangent is $y = -4x + 5$ (using $y = mx + c$)
i.e. $4x + y = 5$

(You should be able to do this for any curve at any point – a standard exam example.)

5·7 how to find the points on a curve where the gradient has a given value.

Example

At which point on the curve with equation $y = 2x^{\frac{3}{2}}$ is the gradient 9?

(Solution) This question becomes, for what values of x is $y'(x) = 9$?

$y = 2x^{\frac{3}{2}} \Rightarrow y'(x) = 3x^{\frac{1}{2}}$;

$y'(x) = 9 \Rightarrow 3x^{\frac{1}{2}} = 9 \Rightarrow \sqrt{x} = 3 \Rightarrow x = 9$;

$x = 9 \Rightarrow y = 2(9)^{\frac{3}{2}} = 2(3^2)^{\frac{3}{2}} = 2(3)^3 = 54$

hence $(9, 54)$

5·8 how to determine whether a curve is increasing, decreasing, or stationary at a given point and in an interval

An increasing curve is going up the page from left to right, the gradient is positive, $y' > 0$.
A decreasing curve goes down the page from left to right, the gradient is negative, $y' < 0$.
A curve is stationary when the tangent is parallel to the x-axis, the gradient is zero.

Example

a) Determine whether the curve with equation $y = x + \frac{1}{x}$ is increasing, decreasing or stationary at i) $x = \frac{1}{2}$ ii) $x = 1$ iii) $x = 2$

b) Determine the range of values of x for which this curve is increasing.

(Solution) a) (First find the derivative) $y = x + \frac{1}{x} = x + x^{-1} \Rightarrow \frac{dy}{dx} = 1 - x^{-2} = 1 - \frac{1}{x^2}$

i) at $x = \frac{1}{2}$, $y' = 1 - 4 = -3$ (<0, so decreasing)

ii) at $x = 1$, $y' = 0$, so stationary

iii) at $x = 2$, $y' = 1 - \frac{1}{4} = \frac{3}{4}$ (>0, so increasing)

If $f'(x) > 0$ for all x in an interval, f is said to be strictly increasing in that interval.
If $f'(x) < 0$ for all x in an interval, f is said to be strictly decreasing in that interval.

b) If y is increasing, then $y' > 0$ i.e. $1 - \frac{1}{x^2} > 0$

Since x^2 is always positive, we can multiply through this inequality by x^2.

$\Rightarrow x^2 - 1 > 0 \Rightarrow x^2 > 1$

$\Rightarrow y$ is increasing for x, $\{x \in \mathbb{R} : x < -1\} \cup \{x \in \mathbb{R} : x > 1\}$

[The simple answer '$x < -1$, $x > 1$' would be acceptable.]

[We can ignore the discontinuity at $x = 0$ as it is not within our answer]

5·9 how to find the stationary points on a curve and determine their nature (i.e. whether maximum or minimum turning point, or point of inflection)

Example

Find the stationary points on the curve with equation $y = x^3 - 3x$.

(Solution) Stationary points occur where $y' = 0$. There is usually a mark for stating this explicitly. (Do not throw this mark away – it is probably the easiest mark in the paper!)

$y' = 3x^2 - 3 = 3(x^2 - 1) = 3(x - 1)(x + 1)$;

for stationary points $y' = 0 \Rightarrow x = -1, 1$

$\Rightarrow y = 2, -2 \Rightarrow (-1, 2)$ and $(1, -2)$

(We can determine the nature of each of these by considering the sign of $\frac{dy}{dx}$ in the neighbourhood of each and gathering these results in a nature table.

x	-1^-	-1	-1^+	1^-	1	1^+
$(x+1)$	$-$	0	$+$	$+$	$+$	$+$
$(x-1)$	$-$	$-$	$-$	$-$	0	$+$
$3(x+1)(x-1) = y'$	$+$	0	$-$	$-$	0	$+$
tangent	↗	→	↘	↘	→	↗

Hence $(-1, 2)$ is a maximum, $(1, -2)$ is a minimum turning point

(Always show clearly in the second last row of the nature table the function that you are using for y' in order to justify the signs written in that row.)

Hints and Tips

An alternative to the nature table in dealing with stationary points is to use the second derivative. This is not listed as part of the syllabus, but there is nothing to stop you using it where you can.

$y'' = 6x \Rightarrow y''(-1) = -6 < 0 \Rightarrow (-1, 2)$ is a maximum turning point.

$y''(1) = 6 > 0 \quad \Rightarrow (1, -2)$ is a minimum turning point

You can learn these rules:
the second derivative is negative at a maximum turning point
the second derivative is positive at a minimum turning point

Hints and Tips continued ➢

Hints and Tips continued

Very able candidates will hopefully understand why. At a maximum turning point for example, as x increases, the curve changes from increasing to stationary to decreasing,

i.e. the gradient goes from positive to zero to negative,
or y' goes from positive to zero to negative,

so y' is decreasing, hence its derivative is negative, i.e. y'' is negative.

5·10 how to sketch a curve with a given equation and interpret the resulting graph

In addition to finding stationary points and their nature, we also need to find intersections with axes, any symmetry, and behaviour of the curve for large positive or negative values of x.

Example

a) Sketch the curve with equation $y = x^3 - 3x$.

b) Use your sketch to determine the values of k for which the equation $x^3 - 3x = k$ has exactly one real root. [This part is of greater difficulty than grade C.]

(Solution) a) intersections with axes:

$x = 0 \Rightarrow y = 0$

$y = 0 \Rightarrow x^3 - 3x = 0 \Rightarrow x(x^2 - 3) = 0 \Rightarrow x = 0, \pm\sqrt{3}$;

hence $(0, 0)$, $(-\sqrt{3}, 0)$, $(\sqrt{3}, 0)$

stationary points:

This is the same function as item 5·9, so we know that $(-1, 2)$ is a maximum turning point and $(1, -2)$ is a minimum turning point

symmetry:

It is not essential, but it is useful, to notice that $x^3 - 3x$ is an odd function, as all the indices are odd numbers. Thus if we replace each x by $(-x)$ we obtain

$y(-x) = (-x)^3 - 3(-x) = -x^3 + 3x = -(x^3 - 3x) = -y$

So if (x, y) lies on the curve, so does $(-x, -y)$.

Hence the curve has half turn symmetry about the origin.
[The turning points have already been shown to satisfy this condition.]
An important part of this type of question is ensuring that all the conclusions you make are consistent with each other. This is also the main purpose of the next (optional) step.

Example continued ➢

***Example** continued*

behaviour for large x (positive and negative)

The leading term is x^3, so that when x is large the graph behaves in a similar way to x^3. This tells us that
as $x \to +\infty$, $y \to (+\infty)^3 \to +\infty \Rightarrow$ graph leaves the page at top right
as $x \to -\infty$, $y \to (-\infty)^3 \to -\infty \Rightarrow$ graph enters the page at bottom left

Putting all this information together gives this sketch:
Make sure you show the 'tails' going beyond the turning points.

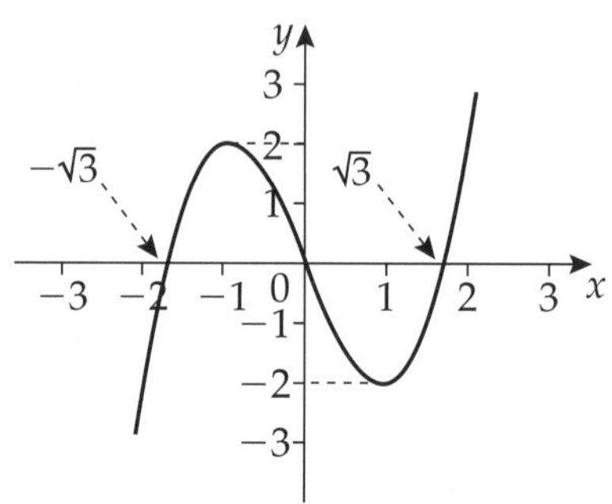

b) The solutions of the equation $x^3 - 3x = k$ must be the simultaneous solutions of $y = x^3 - 3x$ and $y = k$. For one real root we look for when $y = k$ crosses $y = x^3 - 3x$ once. (The sketch graph shows three crossing points.) From the graph, clearly the answer is $\{k < -2\} \cup \{k > 2\}$.

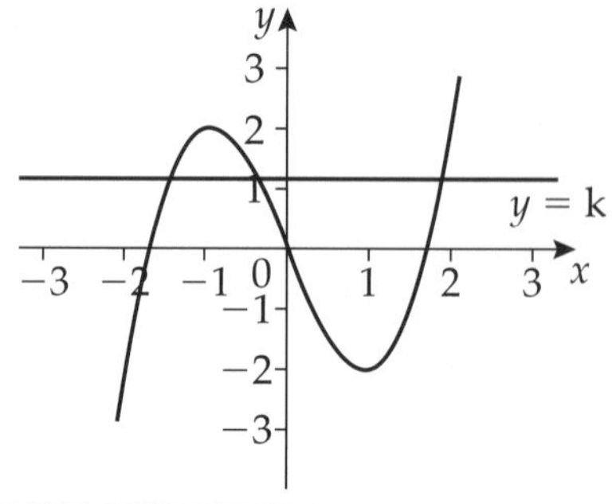

5·11 how to determine the greatest and least values of $y = f(x)$ on the interval $a \leqslant x \leqslant b$

To answer such a question, you generally sketch the curve $y = f(x)$, draw the lines $x = a$ and $x = b$, and examine the curve between these lines. [Some teachers would tell you to shorten your working by ignoring the stationary points which fall outside the interval $a \leqslant x \leqslant b$.] The crux of the matter is that the greatest and least values may occur either at a turning point (a local maximum or local minimum) or at a boundary.

Example

Find the greatest and least values of $x^3 - 3x$ in the interval $\frac{1}{2} \leqslant x \leqslant \frac{3}{2}$.

(Solution) This is the graph which we have just drawn in item **5·10**.

Add the lines $x = \frac{1}{2}$ and $x = \frac{3}{2}$ and evaluate y at these boundaries

$$x = \frac{1}{2} \Rightarrow y = \frac{1}{8} - \frac{3}{2} = -\frac{11}{8}$$

$$x = \frac{3}{2} \Rightarrow y = \frac{27}{8} - \frac{9}{2} = -\frac{9}{8}$$

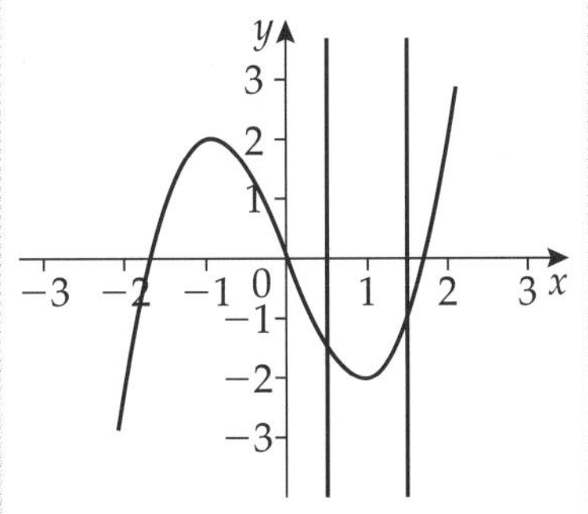

consider the minimum turning point $(1, -2)$

Thus for $\frac{1}{2} \leqslant x \leqslant \frac{3}{2}$, $-2 \leqslant y \leqslant -\frac{9}{8}$.

5·12 how to solve problems requiring optimisation

(See Chapter 18 on typical longer exam questions: Questions **18·2**, **18·11**, **18·15** and **18·16**.)

Hints and Tips

You must learn to take responsibility for your own learning.
This is at the heart of being well motivated.

Chapter 6

RECURRENCE RELATIONS

What You Should Know

6·1 that a linear recurrence relation is defined by $u_{n+1} = au_n + b$ and a value for u_0, where u_n denotes the nth term of a sequence

6·2 how to evaluate successive terms of a given recurrence relation

6·3 that when $|a| < 1$ (i.e. $-1 < a < 1$) for the above recurrence relation, as $n \to \infty$, u_n tends to a limit, and the value of the limit is independent of the value of u_0

6·4 how to evaluate the limit, l, of such a sequence $\left(\text{either using } l = al + b \text{ or } l = \dfrac{b}{1-a}\right)$

6·5 how to apply the above in context

Notes on items 6·1 to 6·5: Recurrence Relations

6·1 that a linear recurrence relation is defined by $u_{n+1} = au_n + b$ and a value for u_0, where u_n denotes the nth term of a sequence

Consider the sequence of the multiples of 5, namely 5, 10, 15, 20, 25, 30, 35, 40, …

This sequence can be defined completely by the formula for the nth term in terms of n, namely $u_n = 5n$. From this any term can be found, e.g. $u_{100} = 500$.

This sequence can also be defined by the recurrence relation $u_{n+1} = u_n + 5$ together with the starting value $u_0 = 0$. From this we can find

$$u_1 = u_0 + 5 = 0 + 5 = 5$$
$$u_2 = u_1 + 5 = 5 + 5 = 10$$
$$u_3 = u_2 + 5 = 10 + 5 = 15 \ldots$$

If we wish to use the recurrence relation to find the 100th term, we need to evaluate all of the first hundred terms in turn.

6·2 how to evaluate successive terms of a given recurrence relation

Example

Find the first five terms of the sequence defined by the recurrence relation $u_{n+1} = 3u_n + 2$, with $u_0 = 4$.

(Solution) The working for this is just the same as I showed you in 6·1, but it is so basic that in exams many people forget what to do because they have more recently been using recurrence relations in more complicated ways.

$$u_0 = 4 \Rightarrow u_1 = 3u_0 + 2 = 3 \times 4 + 2 = 14$$
$$\Rightarrow u_2 = 3u_1 + 2 = 3 \times 14 + 2 = 44$$
$$\Rightarrow u_3 = 3u_2 + 2 = 3 \times 44 + 2 = 134$$
$$\Rightarrow u_4 = 3u_3 + 2 = 3 \times 134 + 2 = 404$$
$$\Rightarrow u_5 = 3u_4 + 2 = 3 \times 404 + 2 = 1214$$

6·3 **that when $|a| < 1$ (i.e. $-1 < a < 1$) for the above recurrence relation, as $n \to \infty$, u_n tends to a limit, and the value of the limit is independent of the value of u_0.**

Before finding the limit of a sequence, remember to justify that it is valid to do so by WRITING DOWN that a limit exists because $|a| < 1$, quoting the value of a that is relevant, not just a general 'a'. See item 6·4.

6·4 **how to evaluate the limit l of such a sequence $\left(\text{either using } l = al + b \text{ or } l = \frac{b}{1-a}\right)$**

Example

Discuss the behaviour of the sequence defined by the recurrence relation $u_{n+1} = 0{\cdot}8u_n + 3$ (with $u_0 = 5$), as $n \to \infty$.

(Solution) Justification of the existence of a limit: (this is usually worth a mark)
$-1 < 0{\cdot}8 < 1 \Rightarrow$ there is a limit, call it L, then

Method 1

$L = 0{\cdot}8L + 3$

$\Rightarrow 0{\cdot}2L = 3$

$\Rightarrow L = \frac{3}{0{\cdot}2} = \frac{30}{2}$ *

$\Rightarrow L = 15$

Method 2

using $l = \frac{b}{1-a}$

$\Rightarrow L = \frac{3}{1 - 0{\cdot}8}$

$\Rightarrow L = \frac{3}{0{\cdot}2} = \frac{30}{2}$ *

$\Rightarrow L = 15$

* Notice how to deal with a vulgar fraction which includes a decimal—multiply top and bottom by 10.
(You must use one of these algebraic methods to answer a question like this. You will not be given any credit for numerous pushes on the buttons of your calculator. You could have done that in primary school. Furthermore, if the limit is say $\frac{6}{7}$, then the calculator will only show a decimal approximation to that.)

6·5 **how to apply the above in context**

(See Chapter 18 on typical longer exam questions: Question **18·3**.)

Remember

Attendance at class is the most vital ingredient of any successful mathematical education.

It is the cumulative effect of all your daily efforts that go to producing a good performance in the actual exam.

Chapter 7

IDENTITIES AND RADIANS

What You Should Know

7·1 that $\sin^2 x + \cos^2 x = 1$ and $\dfrac{\sin x}{\cos x} = \tan x$

7·2 these exact values of sin, cos and tan of 0°, 30°, 45°, 60°, 90°

	0°	30°	45°	60°	90°
sin	0	$\frac{1}{2}$	$\frac{1}{\sqrt{2}}$	$\frac{\sqrt{3}}{2}$	1
cos	1	$\frac{\sqrt{3}}{2}$	$\frac{1}{\sqrt{2}}$	$\frac{1}{2}$	0
tan	0	$\frac{1}{\sqrt{3}}$	1	$\sqrt{3}$	∞

7·3 how to simplify sin, cos, tan of $(90° - x)$, $(180° - x)$, $(360° - x)$

7·4 how to calculate lengths and angles in three dimensional situations

7·5 that a **radian** is the size of the angle subtended at the centre of any circle by an arc equal in length to the radius, and hence π radians = 180 degrees

7·6 how to convert measures given in radians to degrees, and vice versa

7·7 how to solve linear trigonometrical equations, e.g. $2\cos x + 1 = 0$ for $0 \leqslant x \leqslant 2\pi$

7·8 how to solve quadratic trigonometrical equations by factorising or using the quadratic formula e.g. $\cos^2 x + 2\cos x - 1 = 0$, $0 \leqslant x \leqslant 2\pi$.

7·9 how to solve linear trigonometrical equations with multiple angles, e.g. $2\cos 3x = \sqrt{3}$

7·10 the general features of the graphs of $f(x) = \sin(ax + b)$, $f(x) = \cos(ax + b)$, their amplitudes and periods

7·11 how to draw trigonometrical graphs expressed in degrees or radians

Notes on items 7·1 to 7·11: Identities and Radians

7·1 that $\sin^2 x + \cos^2 x = 1$ and $\dfrac{\sin x}{\cos x} = \tan x$

- These are called identities because they are true for all values of x.
- $\sin^2 x + \cos^2 x = 1$ is a direct consequence of the theorem of Pythagoras and must also be known in its other two forms, namely $\sin^2 x = 1 - \cos^2 x$, and $\cos^2 x = 1 - \sin^2 x$.
- Sometimes you will have to simplify $\dfrac{\sin x}{\cos x}$ to obtain $\tan x$. Other times the key to solving a problem will be to replace $\tan x$ by $\dfrac{\sin x}{\cos x}$.

7·2 these exact values of sin, cos, and tan of 0°, 30°, 45°, 60°, 90°

	0°	30°	45°	60°	90°
sin	0	$\frac{1}{2}$	$\frac{1}{\sqrt{2}}$	$\frac{\sqrt{3}}{2}$	1
cos	1	$\frac{\sqrt{3}}{2}$	$\frac{1}{\sqrt{2}}$	$\frac{1}{2}$	0
tan	0	$\frac{1}{\sqrt{3}}$	1	$\sqrt{3}$	∞

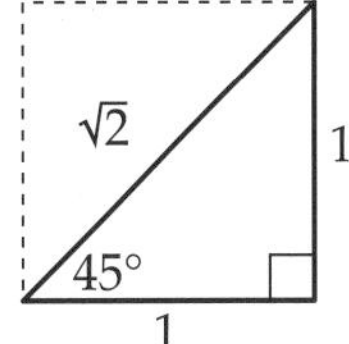

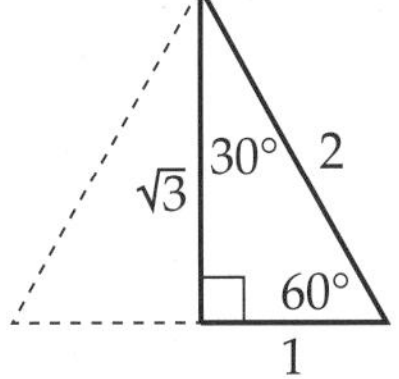

These facts are very likely to be tested in the non-calculator paper. You can easily deduce these results from these diagrams of a half square and half of an equilateral triangle. For 0° and 90°, recall the graphs of sin, cos and tan.

7·3 how to simplify sin, cos, tan of $(90° - x)$, $(180° - x)$, $(360° - x)$

Some of these rules are obvious graphically, making use of the content of item 4·8. The graph of $y = \sin x$ has half turn symmetry about the origin so $\sin(-x) = -\sin(x)$. The graph of $y = \cos x$ is symmetrical about the y-axis so $\cos(-x) = \cos(x)$.

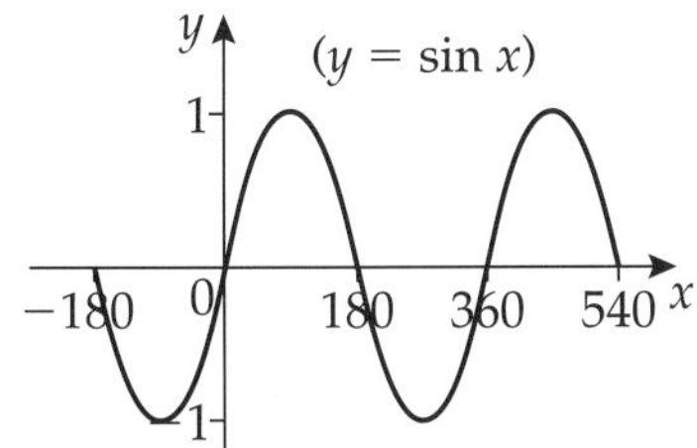

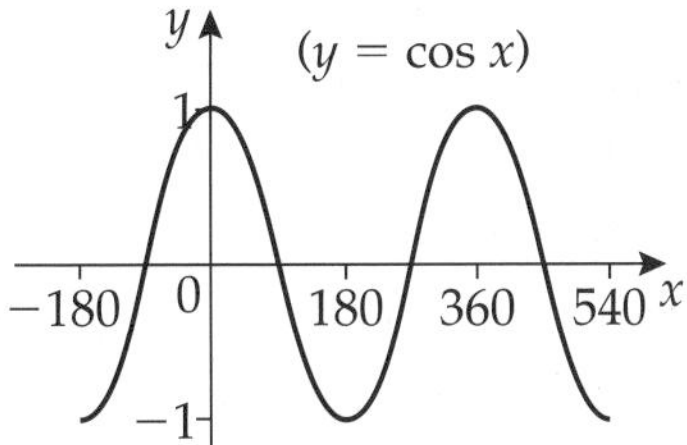

We can make use of the first result to begin to examine $\sin(180 - x)°$.
$\sin(180 - x)° = \sin[-(x - 180)]° = -\sin(x - 180)°$ but $y = -\sin(x - 180)°$ means translate $y = \sin(x)°$ by 180° to the right then reflect in the x-axis. From the graph, it can be seen that the sine graph maps to itself under this composite transformation, hence $\sin(180 - x)° = \sin(x)°$.

We will assume that the following results we obtain for an acute angle A are valid for all values of x.

The trig functions of $(90° - A)$ can be obtained by considering a right angled triangle:

$$\sin(90° - A) = \frac{q}{h} = \cos A$$

$$\cos(90° - A) = \frac{p}{h} = \sin A$$

$$\tan(90° - A) = \frac{q}{p} = \frac{1}{\frac{p}{q}} = \frac{1}{\tan A}$$

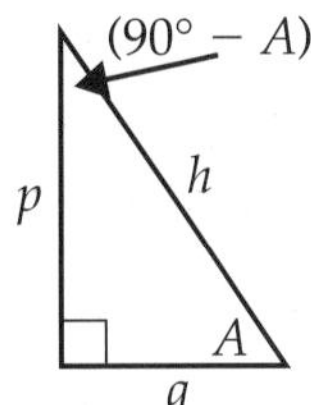

The trig functions of $(180° - A)$ and $(360° - A)$: make use of the proper mathematical definitions of sin, cos, and tan, which apply to all sizes of angle even though I am only showing you an acute case:

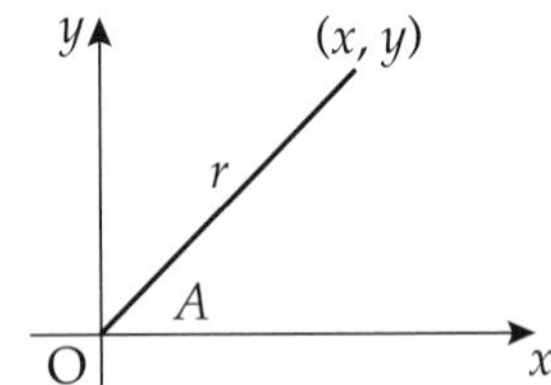

$$\sin(A) = \frac{y}{r} \quad \cos(A) = \frac{x}{r} \quad \tan(A) = \frac{y}{x}$$

For $(180° - A)$: [Also look back at item **4·8**.]

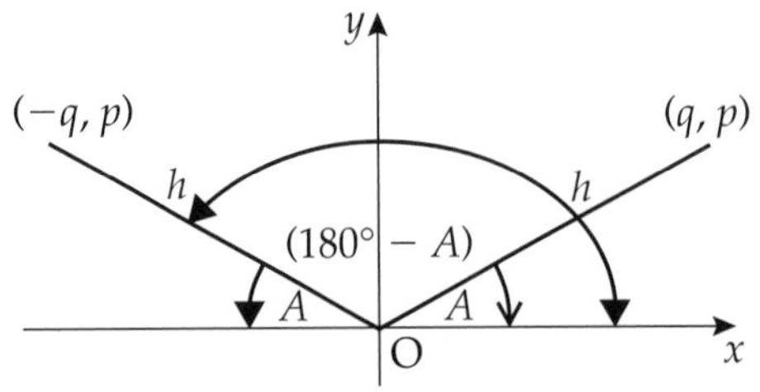

$$\sin(180° - A) = \frac{p}{h} = \sin(A)$$

$$\cos(180° - A) = \frac{-q}{h} = -\cos(A)$$

$$\tan(180° - A) = \frac{p}{-q} = -\tan(A)$$

I leave the results for $(360° - A)$ as an exercise for you to try.
If you also do it for $(180° + A)$, you will see where the 'all sin tan cos' rule comes from.
You can also obtain these results from the compound angle formulae in item **11·1**.

7·4 how to calculate lengths and angles in three dimensional situations.

Most questions here concern calculating:

i) the angle between a line and a plane 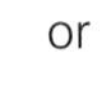 or ii) the angle between two planes.

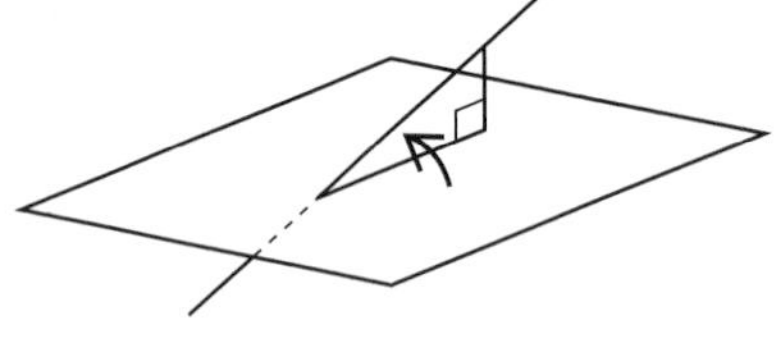

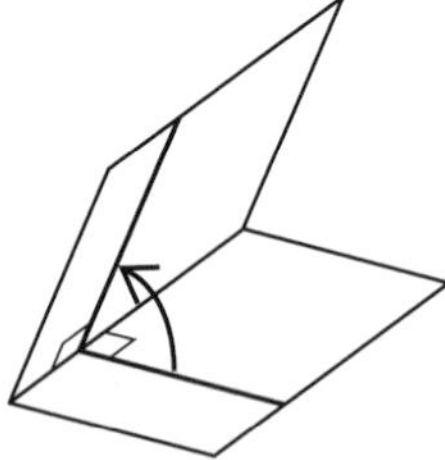

You can normally rely on the theorem of Pythagoras and basic trigonometry. Occasionally you may need the sine rule or the cosine rule.

Example

V, ABCD is a right rectangular pyramid with AB = 8, BC = 6 and AV = 13.

Calculate the angle between

a) AV and the base ABCD

b) the plane VBC and the base ABCD.

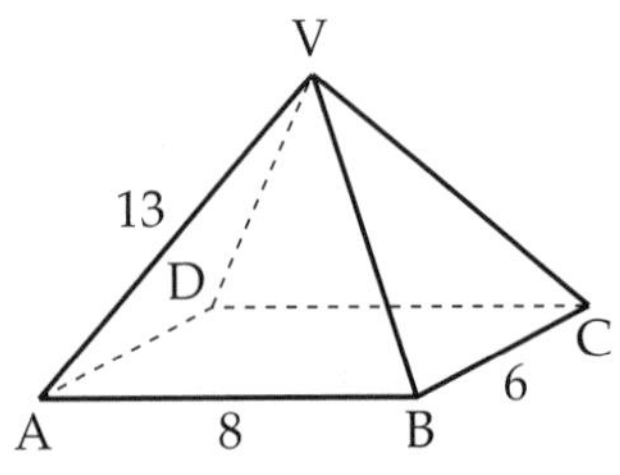

***Example** continued* ➢

Example continued

(Solution) **a)** Let AC and DB cross at E.
AE is the projection of AV on the base, so the angle between AV and the base is $V\hat{A}E$.

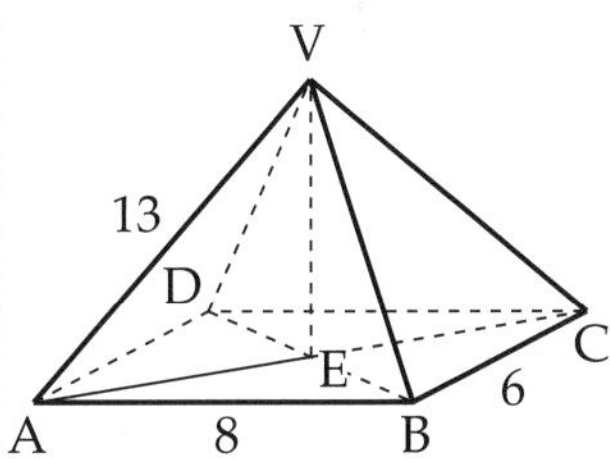

(We only know the length of one side in ΔVAE, so we must find another; AE is the easier. Draw an auxiliary diagram of ΔABC with the right angle at B more obvious.)

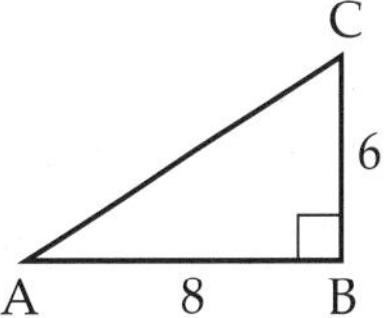

Clearly, by the theorem of Pythagoras,
AC = 10 (a '3, 4, 5' Δ)
$AE = \frac{1}{2}AC = 5$

(Now draw another auxiliary diagram, of ΔVAE this time, and use AE = 5.)

$\therefore \cos(\hat{A}) = \frac{5}{13} \Rightarrow A = 67{\cdot}4°$

Also for part b), we need VE = 12 (a '5, 12, 13' Δ)

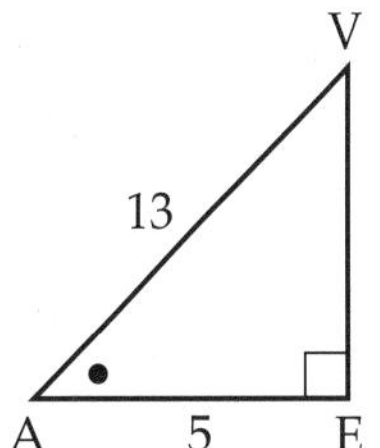

b) The planes meet in the line BC which has a mid point, call it M.
ME lies in the plane ABCD and is perpendicular to BC.
MV lies in the plane VBC and is perpendicular to BC.
So the angle between the planes is $V\hat{M}E$.
[$V\hat{C}E$ and $V\hat{B}E$ are incorrect.]
(Now draw an auxiliary diagram of ΔVME.)

V
13
D
6
C
E
M
A
8
B

$EM = \frac{1}{2}AB = 4.$

$\tan(V\hat{M}E) = \frac{12}{4} = 3$

$\Rightarrow V\hat{M}E = 71{\cdot}6°$

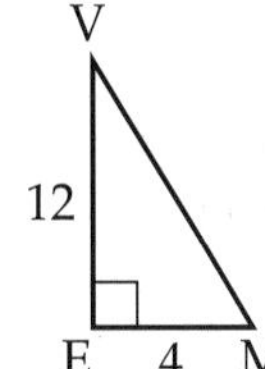

7·5 **that a radian is the size of the angle subtended at the centre of any circle by an arc equal in length to the radius, and hence π radians = 180 degrees.**

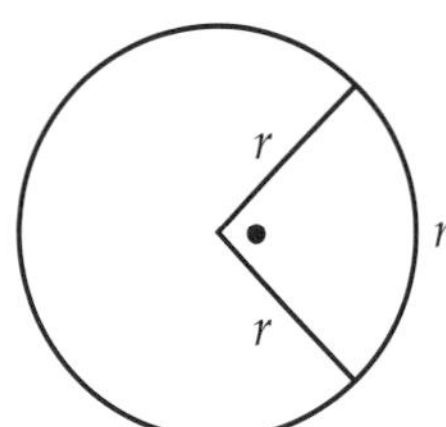

The angle at the centre of a sector is proportional to the length of the arc $\Rightarrow \frac{1^c}{360°} = \frac{r}{2\pi r}$

hence $2\pi^c = 360°$ or $\pi^c = 180°$

The little ^{c} denotes radians (also called circular measure). It is seldom used and is only needed here because we have both units in the one equation. It is shorter than writing radians in full.

It is a standard convention that if there is no degree sign, then radians are intended.

7·6 **how to convert measures given in radians to degrees, and vice versa.**

This is done by direct proportion using π radians = 180°

Example

Express a) $\frac{5\pi}{6}$ radians in degrees b) 300° in radians.

(Solution) a) π radians = 180°

$\Rightarrow \frac{5\pi}{6}$ radians $= \frac{5}{6} \times 180° = 150°$

b) 180° = π radians

$\Rightarrow 300° = 300 \times \frac{\pi}{180} = \frac{10\pi}{6} = \frac{5\pi}{3}$ radians

At this point, if you have not already done so, you should make yourself a copy of the table in 7·2 with the angles in degrees replaced by their equivalents in radians.

You should become comfortable with the common angles $\frac{\pi}{6}, \frac{\pi}{4}, \frac{\pi}{3}, \frac{\pi}{2}$, and multiples of them.

7·7 how to solve linear trigonometrical equations

This item is important because the next item depends directly upon it.

Example

Solve $2\cos x + 1 = 0$ for $0 \leqslant x \leqslant 2\pi$

(Solution) First make $\cos x$ the subject of the equation, taking care not to introduce any silly errors. $\cos x = -\frac{1}{2}$

Now think 'in which quadrants is cos negative?'

Using

sin	All
tan	cos

we get 2nd and 3rd

in the 2nd quadrant we need (π − the acute angle)
in the 3rd quadrant we need (π + the acute angle)

Now ask 'which acute angle has a cosine of $\frac{1}{2}$?'

answer: $\cos\left(\frac{\pi}{3}\right) = \frac{1}{2}$ hence $x = \pi - \frac{\pi}{3}, \pi + \frac{\pi}{3}$

$$\Rightarrow x = \frac{2\pi}{3}, \frac{4\pi}{3}$$

7·8 how to solve quadratic trigonometrical equations by factorising or using the quadratic formula

Example

Solve $\cos^2 x + 2\cos x - 1 = 0$ for $0 \leqslant x \leqslant 2\pi$.

(Solution) Some people find it easier to write this quadratic equation as $C^2 + 2C - 1 = 0$.

(In the exam, you are more likely to be given an example which factorises, but this one doesn't. You can determine this before you try to factorise by calculating the discriminant (see item **8·3**). Since the discriminant in this case is 8, which is not a perfect square, we need the quadratic formula.)

$$C = \frac{-b \pm \sqrt{b^2 - 4ac}}{2a} = \frac{-2 \pm \sqrt{2^2 - 4 \times 1 \times (-1)}}{2 \times 1} = \frac{-2 \pm \sqrt{8}}{2} = -1 \pm \sqrt{2}$$

i.e. $\cos x = -1 - \sqrt{2}$ or $\cos x = -1 + \sqrt{2} = 0{\cdot}414$

but $-1 - \sqrt{2} < -1$ so no roots for x $\Rightarrow$ $x = 1{\cdot}144, 5{\cdot}139$

(Remember to set your calculator in 'radian' mode here.)

7·9 how to solve linear trigonometrical equations with multiple angles

Example

Solve $2\cos 3x° = \sqrt{3}$ for $0 \leq x \leq 360$.

(Solution) $2\cos 3x° = \sqrt{3} \Rightarrow \cos 3x° = \frac{\sqrt{3}}{2}$

$\Rightarrow \quad 3x = 30, 330, 390, 690, 750, 1050$ *

$\Rightarrow \quad x = 10, 110, 130, 230, 250, 350$

*** If we only take the solutions for $3x$ up to 360 at this stage, then we will end up with only the solutions for x which are less than 120. To obtain all the solutions for x up to 360, we must consider all the solutions for $3x$ up to 3×360.**

7·10 the general features of the graphs of $f(x) = \sin(ax + b)$, $f(x) = \cos(ax + b)$, their amplitudes and periods

Example

Sketch the graphs of a) $f(x) = \sin(2x + 30)°$ for $0 \leq x \leq 360$
b) $g(x) = 3\cos\left(\frac{1}{2}x - \frac{\pi}{6}\right)$ for $0 \leq x \leq 4\pi$

(Solution) a) Write this as $f(x) = \sin[2(x + 15)°]$;
the **2** means two complete sine waves between 0 and 360,
i.e. the period of f is 180°
the **+15°** translates (slides) the graph 15° left
$f(x) = 1 \times \sin(\ldots)$ means that the amplitude of the function is 1, hence the max. value of 1 and min. value of −1 remain unaltered

$y = \sin x°$ has a maximum of 1 when $x = 90$, so
$f(x)_{max} = 1$ when $(2x + 30) = 90, 450$ [consider two cycles because of the 2]
$\Rightarrow 2x = 60, 420 \Rightarrow x = 30, 210$

$y = \sin x°$ has a minimum of −1 when $x = 270$, so
$f(x)_{min} = -1$ when $(2x + 30) = 270, 630$
$\Rightarrow 2x = 240, 600 \Rightarrow x = 120, 300$

$y = \sin x°$ has zeros when $x = 0, 180, 360$, so
$f(x) = 0$ when $(2x + 30) = 180, 360$
$\Rightarrow 2x = 150, 330, 510, 690 \Rightarrow x = 75, 165, 255, 345$
for the end points $x = 0, 360 \Rightarrow f(x) = \sin 30° = 0.5$

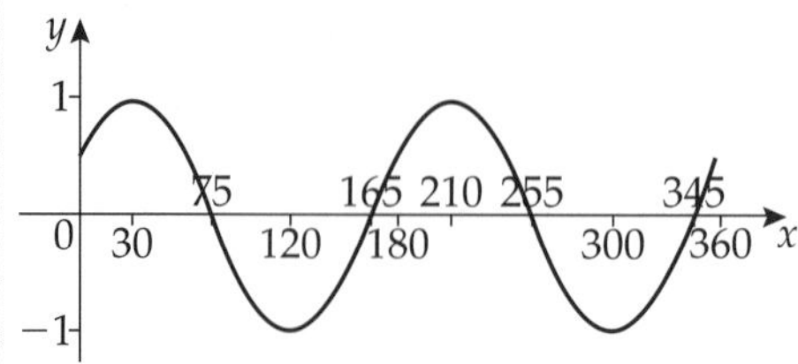

Example continued ➢

Example continued

b) Write this as $g(x) = 3\cos\left[\frac{1}{2}\left(x - \frac{\pi}{3}\right)\right]$

$\cos\left(\frac{1}{2}x\right)$ needs 4π for one complete cosine wave, i.e. the period of g is 4π

the $-\frac{\pi}{3}$ translates the graph $\frac{\pi}{3}$ to the right

the amplitude **3** stretches the graph by a factor of 3 parallel to the y-axis.

$y = \cos x$ has a maximum of 1 at $x = 0, 2\pi$, so

$g(x) = 3$ when $\left[\frac{1}{2}\left(x - \frac{\pi}{3}\right)\right] = 0 \Rightarrow x = \frac{\pi}{3}$

$y = \cos x$ has a minimum of -1 at $x = \pi$, so

$g(x) = -3$ when $\left[\frac{1}{2}\left(x - \frac{\pi}{3}\right)\right] = \pi \Rightarrow x = 2\pi + \frac{\pi}{3} = \frac{7\pi}{3}$

$y = \cos x$ has roots at $x = \frac{\pi}{2}, \frac{3\pi}{2}$, so

$g(x) = 0$ when $\left[\frac{1}{2}\left(x - \frac{\pi}{3}\right)\right] = \frac{\pi}{2}, \frac{3\pi}{2} \Rightarrow x = \pi + \frac{\pi}{3}, 3\pi + \frac{\pi}{3} = \frac{4\pi}{3}, \frac{10\pi}{3}$

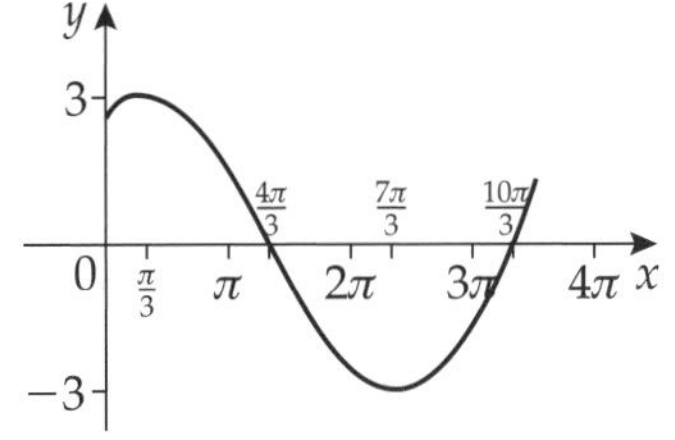

7·11 how to draw trigonmetrical graphs expressed in degrees or radians

Example

Sketch the curve with equation $y = 3\sin\left(\frac{\pi t}{4}\right)$ for $0 \leqslant t \leqslant 8$.

(Solution) Identify the salient features:

max. 3 when, $\sin\left(\frac{\pi t}{4}\right) = 1$ i.e. $\frac{\pi t}{4} = \frac{\pi}{2} \Rightarrow t = 2$

min. -3 when $\sin\left(\frac{\pi t}{4}\right) = -1$ i.e. $\frac{\pi t}{4} = \frac{3\pi}{2} \Rightarrow t = 6$

zeros of y occur when $\sin\left(\frac{\pi t}{4}\right) = 0$

i.e. when $\frac{\pi t}{4} = 0, \pi, 2\pi \Rightarrow t = 0, 4, 8$

putting this together gives

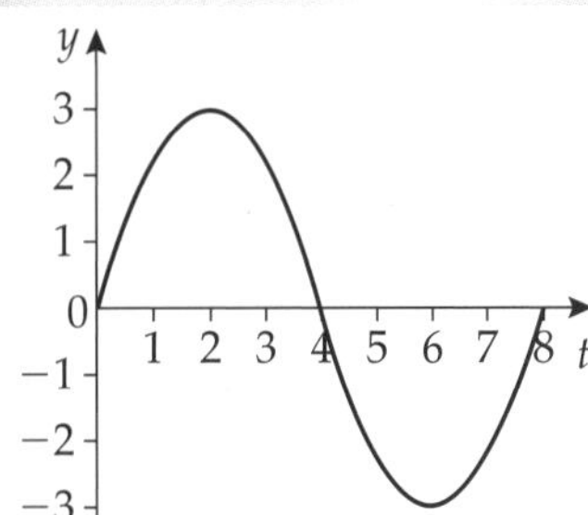

Chapter 8

QUADRATIC THEORY

What You Should Know

8·1 how to solve by factorising, a quadratic equation and a quadratic inequality

8·2 the roots of $ax^2 + bx + c = 0$ are $\dfrac{-b \pm \sqrt{b^2 - 4ac}}{2a}$

8·3 that the discriminant of $ax^2 + bx + c$ is $b^2 - 4ac$, and how it is used to determine whether the roots of a quadratic equation are real or unreal, and, if real, equal or unequal

8·4 how to determine, for a given quadratic equation with some general coefficients, a condition to be satisfied by the coefficients in order for the roots to be real, unreal, equal, unequal

8·5 how to form an equation with given roots

8·6 how to express $\dfrac{a}{b \pm \sqrt{c}}$ with a rational denominator

8·7 that when a linear and second degree equation are solved simultaneously, equal roots of the resulting quadratic equation indicate that the line is a tangent to the curve

Notes on items 8·1 to 8·7: Quadratic Theory

8·1 how to solve by factorising, a quadratic equation and a quadratic inequality

Solving a quadratic equation by factorising is revision and was covered in item 2·5. We still need to look at an example of a quadratic inequality.

Example

Solve $3 + 2x - x^2 \geqslant 0$.

(Solution) Factorising gives $(3 - x)(1 + x) \geqslant 0$
The roots of this quadratic function are -1 and 3.
The '$-x^2$' indicates a maximum turning point.
Sketch the graph of $y = 3 + 2x - x^2$.
(This was covered in item 2·6.)

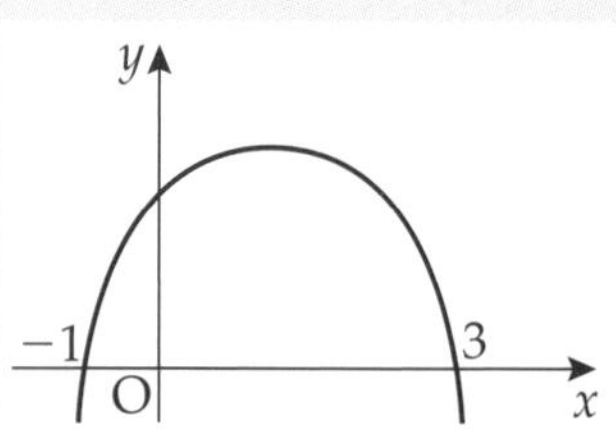

(We do not need the coordinates of the turning point in this context.)
The function is positive when its graph is above the x-axis.
hence the solution is $-1 \leqslant x \leqslant 3$

8·2 **the roots of $ax^2 + bx + c = 0$ are $x = \dfrac{-b \pm \sqrt{b^2 - 4ac}}{2a}$**

(This is revision. An example of its use has previously been given in item **2·5**.)

8·3 **that the discriminant of $ax^2 + bx + c$ is $b^2 - 4ac$, and how it is used to determine whether the roots of a quadratic equation are real or unreal, and, if real, equal or unequal**

In the equation $ax^2 + bx + c = 0$, the expression $ax^2 + bx + c$ has discriminant $\Delta = b^2 - 4ac$, and

- $\Delta < 0 \Rightarrow$ no real roots
- $\Delta = 0 \Rightarrow$ equal real roots $\left(\text{moreover they must be rational, each is } \dfrac{-b}{2a}\right)$
- $\Delta > 0 \Rightarrow$ distinct real roots (distinct means unequal)

Note:

1. **Δ is that part of the quadratic formula under the square root sign**
2. **real roots occur when $\Delta \geqslant 0$ (be careful about this; do not confuse $>$ with $\geqslant$)**
3. **when $\Delta \geqslant 0$ and Δ is also a perfect square, then $ax^2 + bx + c$ will factorise and the roots of the equation will be rational numbers (non-examinable but very useful)**
4. **when $\Delta \geqslant 0$ but Δ is not a perfect square, then $ax^2 + bx + c$ will not have rational factors and the roots will be irrational numbers, (non-examinable but very useful), which should be left as surds if exact answers are required**
5. **remember that the set of real numbers is the union of the set of rational numbers (i.e. fractions) and the set of irrational numbers (e.g. $\sqrt{3}$, π, e)**

Example

Determine the nature of the roots of the equation

a) $2x^2 + 5x - 1 = 0$
b) $x^2 - 4x + 9 = 0$
c) $2x^2 + (p + 2)x + p = 0$

(Solution)

a) Compare $2x^2 + 5x - 1 = 0$ with $ax^2 + bx + c = 0$
in this case $a = 2 \quad b = 5 \quad c = -1$
so $\Delta = b^2 - 4ac = 5^2 - 4 \times 2 \times (-1) = 25 + 8 = 33$ (>0)
$\Rightarrow$ distinct real roots (they will also be irrational)

b) $a = 1 \quad b = -4 \quad c = 9$ (this line may be omitted)
$\Rightarrow \Delta = (-4)^2 - 4 \times 1 \times 9 = 16 - 36 = -20 < 0 \Rightarrow$ no real roots

c) $a = 2 \quad b = (p + 2) \quad c = p$ (note that none of these contains x)

$$\begin{aligned}\Rightarrow \Delta &= (p+2)^2 - 4 \times 2 \times p \\ &= p^2 + 4p + 4 - 8p \\ &= p^2 - 4p + 4 \\ &= (p-2)^2 \ (>0 \text{ for all } p)\end{aligned}$$

$\Rightarrow$ the roots are always real
(in fact they are always rational)

8·4 how to determine, for a given quadratic equation with some general coefficients, a condition to be satisfied by the coefficients in order for the roots to be real, unreal, equal, unequal

Example

For what values of k does $kx^2 - 3(k+1)x + 12 = 0$ have two real roots?

(Solution) $kx^2 - 3(k+1)x + 12 = 0$

For two real roots, the discriminant is strictly greater than zero

using 'a' $= k$, 'b' $= -3(k+1)$, 'c' $= 12$ (**Note:** none of 'a', 'b', 'c' involves x.)

then $\Delta = [-3(k+1)]^2 - 4 \times k \times 12$

$= 9(k^2 + 2k + 1) - 48k = 9k^2 - 30k + 9 = 3[3k^2 - 10k + 3]$

$= 3(3k-1)(k-3)\ (>0)$

this is a quadratic inequality, which was covered in item 8·1.

roots occur at $\frac{1}{3}$ and 3; also $k = 0 \Rightarrow \Delta = 9$

so the sketch looks like this:

$\Delta > 0$ when its graph is above the k-axis, so there are two real roots for $\left\{k < \frac{1}{3}\right\} \cup \{k > 3\}$

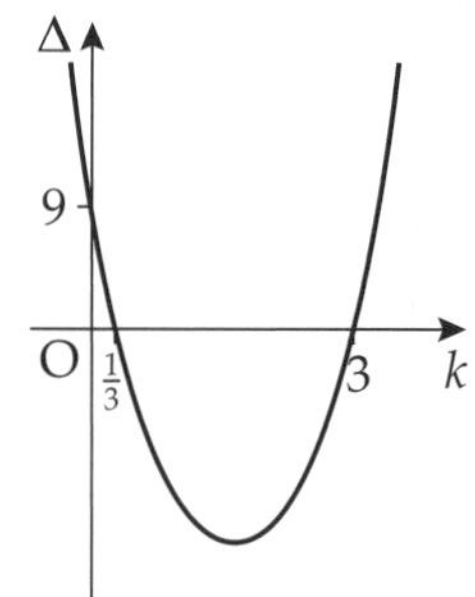

8·5 how to form an equation with given roots

Example

Form an equation with roots -2 and 3.

(Solution) If 3 is a root of $f(x) = 0$, then $(x-3)$ must be a factor of $f(x)$.

similarly if -2 is a root, then $(x+2)$ must be a factor.

Hence a possible equation is $(x+2)(x-3) = 0$ i.e. $x^2 - x - 6 = 0$

(Any multiple of this equation (positive or negative) is also acceptable.)

Note: this process has already been made use of in item 4·11. This is a straightforward type of question which recently has not been done well by Higher Maths candidates.

8·6 how to express $\dfrac{a}{b \pm \sqrt{c}}$ with a rational denominator

We make use of the difference of two squares: $p^2 - q^2 = (p - q)(p + q)$.

Even if p and q are both (quadratic) surds, $p^2 - q^2$ must always be rational.

So if $b - \sqrt{c}$ appears on the bottom line, we know that $(b - \sqrt{c})(b + \sqrt{c}) = b^2 - c$, so we multiply both numerator and denominator by the conjugate surd $b + \sqrt{c}$.

Example

Express $\dfrac{2}{3+\sqrt{5}}$ with a rational denominator.

(Solution) $\dfrac{2}{3+\sqrt{5}} = \dfrac{2}{3+\sqrt{5}} \times \dfrac{3-\sqrt{5}}{3-\sqrt{5}} = \dfrac{2(3-\sqrt{5})}{3^2-(\sqrt{5})^2} = \dfrac{2(3-\sqrt{5})}{4} = \dfrac{1}{2}(3-\sqrt{5})$

8·7 that when a linear and second degree equation are solved simultaneously, equal roots of the resulting quadratic equation indicate that the line is a tangent to the curve

Example

Show that the line with equation $4x - y - 1 = 0$ is a tangent to the parabola with equation $y = 2x^2 + 12x + 7$ and find the point of contact.

(Solution) Solve the system of equations simultaneously. Re-arrange the linear equation (carefully) and substitute it into the quadratic equation. [This method applies to all second degree curves (circles etc.).] It is very common for pupils to make an error with re-arranging the linear equation. Make sure you don't lose a mark on the easy bit – it could make the rest more difficult or even impossible. If you find at the end that you do not have equal roots, then look over your work for an error; it's probably at the beginning. If it is, start again. Scoring out and over writing is usually a recipe for disaster.

$4x - y - 1 = 0 \Rightarrow y = 4x - 1$

replace y in the parabola equation by $4x - 1 \Rightarrow 4x - 1 = 2x^2 + 12x + 7$

simplifying gives $2x^2 + 8x + 8 = 0$

$\Rightarrow x^2 + 4x + 4 = 0$*

$\Rightarrow (x + 2)^2 = 0$

$\Rightarrow x = -2, -2$

equal roots $\Rightarrow$ the line is a tangent to the parabola

[Don't miss the above line out – communicate!]

now use $y = 4x - 1$ to find y, $y = 4 \times (-2) - 1 = -9$ i.e. $(-2, -9)$

(* alternatively at this point you could use the discriminant: $\Delta = 16 - 4 \times 1 \times 4 = 0 \Rightarrow$ equal roots $\Rightarrow$ the line is a tangent to the parabola; but you still have to solve the equation if you want the point of contact.

Chapter 9

THE REMAINDER THEOREM

What You Should Know

9·1 the meaning of the terms polynomial, coefficient, order of a polynomial, power, index, degree

9·2 how to evaluate $f(k)$ (k a constant) where $f(x)$ is any given polynomial (you should also be able to program your calculator to do this task)

9·3 that the remainder on dividing a polynomial $f(x)$ by $(x - h)$ completely is $f(h)$

9·4 that if $f(h) = 0$, then $(x - h)$ is a factor of $f(x)$

9·5 how to factorise cubic and quartic polynomials (with at most one quadratic factor, the others being linear)

9·6 how to solve cubic and quartic equations

9·7 the general features of the graphs of simple polynomial functions e.g. $f(x) = x(x - 1)^2$

9·8 how to prove that an equation has a root between two given (appropriate) values, and find that root to any required degree of accuracy

Notes on items 9·1 to 9·8: The Remainder Theorem

9·1 the meaning of the terms polynomial, coefficient, order of a polynomial, power, index, degree

Key Words and Definitions

In general, a **polynomial** is of the form $f(x) = a_0 + a_1x + a_2x^2 + a_3x^3 + \ldots\ldots + a_nx^n$.
For example $g(x) = 1 + 3x - 5x^2$, $h(x) = 2x^3 - 4x^2 + 5x - 1$.

$g(x)$ is said to be written in **ascending** order and $h(x)$ in **descending** order of terms.

In $f(x)$, $a_0, a_1, a_2, a_3, \ldots\ldots, a_n$ are the **coefficients**.

The coefficient of x^2 in $g(x)$ is -5, and the coefficient of x in $h(x)$ is 5.

The **degree** of the polynomial $g(x)$ is 2 as this is the highest **power** involved.

The degree of $h(x)$ is 3 as this is the greatest **index** appearing.

9·2 how to evaluate $f(k)$ (k a constant) where $f(x)$ is any given polynomial (you should also be able to program your calculator to do this task)

Example Evaluate $f(3)$ where $f(x) = 2x^3 - 4x^2 + 5x - 1$.

(Solution) (There are four methods of doing this. You should know them all and select the most appropriate each time. You can often use a second method as a check on the first one that you use.)

by direct substitution:
$f(3) = 2(3)^3 - 4(3)^2 + 5(3) - 1 = 54 - 36 + 15 - 1 = 32$
(It is useful to remember that f(1) is the sum of the coefficients.)

by programmable calculator:
I am not going to try to explain this because the chances of your calculator working exactly the same way as mine are quite remote.

Use your calculator manual to ensure that you can enter a polynomial and then find $f(1)$, $f(2)$, … etc. successively without further laborious input. You need this for item **9·8**, which involves evaluating $f(2{\cdot}34)$ etc.

by nesting:
$f(x) = 2x^3 - 4x^2 + 5x - 1 = (2x^2 - 4x + 5)x - 1 = ([2x - 4]x + 5)x - 1$
$\Rightarrow f(3) = ([2 \times 3 - 4] \times 3 + 5) \times 3 - 1 = (2 \times 3 + 5) \times 3 - 1 = 11 \times 3 - 1 = 32$

by synthetic division:
The top row consists of the coefficients in descending order.

Remember to include a zero for any 'missing' power. See item **9·4**.

This is really just a different way of laying out the working for the nesting process.

3	2	−4	5	−1
		6	6	33
	2	2	11	32

Copy down the 2. Then follow the first arrow : $2 \times 3 = 6$.
Then: $(-4) + 6 = 2$ Then $2 \times 3 = 6$ $5 + 6 = 11$ $11 \times 3 = 33$ $(-1) + 33 = 32$

9·3 that the remainder on dividing a polynomial $f(x)$ by $(x - h)$ completely is $f(h)$

This is known as the remainder theorem.

Example Find the remainder on dividing $5x^3 - 3x^2 + 7x - 2$ *by* $(x - 2)$

(Solution) Let $f(x) = 5x^3 - 3x^2 + 7x - 2$, then the remainder is $f(2)$.
Remainder $= f(2) = (5 \times 8) - (3 \times 4) + (7 \times 2) - 2 = 54 - 14 = 40$

Alternatively

2	5	−3	7	−2
		10	14	42
	5	7	21	40

(As with many mathematical processes you sometimes also have to work backwards, as in the next example.)

Example

Find the value of k for which $2x^3 + 5x^2 + kx + 7$ has a remainder of 10 when divided by $(x + 3)$.

(Solution) Let $2x^3 + 5x^2 + kx + 7 = f(x)$, then the remainder is $f(-3)$.

Think: 'f(the value of x which makes the divisor zero)'

Method 1

$f(-3) = 10$

$\Rightarrow -54 + 45 - 3k + 7 = 10$

$\Rightarrow -3k = 12$

$\Rightarrow k = -4$

Method 2

$$\begin{array}{r|rrrr} -3 & 2 & 5 & k & 7 \\ & & -6 & 3 & -3k-9 \\ \hline & 2 & -1 & k+3 & -3k-2 \end{array}$$

$\Rightarrow -3k - 2 = 10$

$\Rightarrow k = -4$

Method 3

$$\begin{array}{r|rrrr} -3 & 2 & 5 & k & 7 \\ & & -6 & 3^* & 3 \\ \hline & 2 & -1 & -1^{**} & 10 \end{array}$$

* work forwards to here

** work backwards to here

hence $k + 3 = -1 \Rightarrow k = -4$

9·4 that if $f(h) = 0$, then $(x - h)$ is a factor of $f(x)$

This is a special case of the remainder theorem. If the remainder is zero, then the divisor must be a factor. Some people elevate it to a theorem in its own right and call it the factor theorem. The method is just the same as that shown above.

Example

Show that $(x - 2)$ is the only linear factor of $x^3 + x^2 - 12$.

(Solution) Let $x^3 + x^2 - 12 = f(x)$

$\Rightarrow f(2) = 8 + 4 - 12 = 0 \Rightarrow (x - 2)$ is a factor of $x^3 + x^2 - 12$.

***Example** continued* ➢

Example continued

This can be confirmed by synthetic division:

2	1	1	0	−12	
		2	6	12	
	1	3	6	0	(0 remainder indicates a factor)

Observe now that $x^3 + x^2 - 12 = (x - 2)(x^2 + 3x + 6)$.
Expand the brackets on the right hand side to confirm this.
Note that the coefficients of the quadratic factor are the numbers on the bottom line of the synthetic division.
Now $x^2 + 3x + 6$ has $\Delta = 9 - (4 \times 1 \times 6) = -15\ (<0) \Rightarrow$ no real factors
Hence $(x - 2)$ is the only linear factor of $x^3 + x^2 - 12$.

9·5 how to factorise cubic and quartic polynomials (with at most one quadratic factor, the others being linear)

Example

Factorise $6x^3 - 25x^2 + x + 60$ fully.

(Solution) Let $6x^3 - 25x^2 + x + 60 = f(x)$.
We try to find a root of $f(x)$ by trial.
Any whole number which is a root will be a factor of 60.
$f(1) = 6 - 25 + 1 + 60 = 42 \neq 0$
$f(-1) = -6 - 25 - 1 + 60 = 28 \neq 0$
$f(2) = 48 - 100 + 2 + 60 = 10 \neq 0$
$f(-2) = -48 - 100 - 2 + 60 = -90 \neq 0$
$f(3) = 162 - 225 + 3 + 60 = 0 \Rightarrow (x - 3)$ is a factor of $f(x)$

Some people can find the quadratic factor by inspection or long division, but the most popular method seems to be synthetic division (even for evaluating $f(\pm 1)$, $f(\pm 2)$ etc). [The synthetic division can be a useful check on the working for the previous substitution, if that method is used first.]

3	6	−25	1	60
		18	−21	−60
	6	−7	−20	0

$\Rightarrow f(x) = (x - 3)(6x^2 - 7x - 20)$
$= (x - 3)(2x - 5)(3x + 4)$

Note how the quadratic factor is obtained from the bottom line of the table. Be very careful to extract the correct signs when doing so and avoid another common error.

Example

Factorise $x^4 + 4x^3 + 8x^2 + 8x + 3$ fully. [Degree 4 is called quartic.]

(Solution) The method starts off by identifying one linear factor and finding the appropriate cubic factor to go with it. We then factorise the cubic factor as shown in the previous example. When doing so, remember that there is no reason why the first linear factor of the cubic cannot be the same as the first linear factor that you found for the quartic. In this case it is. Note also that when we are left with a quadratic factor that does not look like it is going to factorise, it is not good enough to state 'This does not factorise'. Prove that it cannot factorise by using the discriminant. Successive synthetic division provides a neat solution to this problem:

$$\begin{array}{r|rrrrr} -1 & 1 & 4 & 8 & 8 & 3 \\ & & -1 & -3 & -5 & -3 \\ \hline -1 & 1 & 3 & 5 & 3 & 0 \\ & & -1 & -2 & -3 & \\ \hline & 1 & 2 & 3 & 0 & \end{array}$$

$\Rightarrow (x + 1)(x^3 + 3x^2 + 5x + 3)$

$\Rightarrow (x + 1)(x^2 + 2x + 3)$

hence $(x + 1)(x + 1)(x^2 + 2x + 3)$
for $x^2 + 2x + 3$, $\Delta = 4 - 4 \times 1 \times 3 = -8\ (<0) \Rightarrow$ no real factors
hence $(x + 1)^2 (x^2 + 2x + 3)$

9·6 how to solve cubic and quartic equations

The most common method of solving quadratic equations is to factorise and obtain the roots by considering each factor in turn to be zero. In principle, this is what we do with higher degree equations as well.

Example

Solve a) $6x^3 - 25x^2 + x + 60 = 0$ b) $x^4 + 4x^3 + 8x^2 + 8x + 3 = 0$.

(Solution) Note that these are the same polynomials as the previous item, so we know their factors.

a) $6x^3 - 25x^2 + x + 60 = 0 \Rightarrow (x - 3)(2x - 5)(3x + 4) = 0$
$\Rightarrow x = 3, \frac{5}{2}, -\frac{4}{3}$

b) $x^4 + 4x^3 + 8x^2 + 8x + 3 = 0 \Rightarrow (x + 1)^2 (x^2 + 2x + 3) = 0$
$\Rightarrow x = -1$ (twice)

(You can see that there is only one more line in the solution of an equation than in factorising a polynomial, but do not add the roots after factorising if factorising is all you have been asked to do. It will not be penalised because it is working following a correct answer, but **show that you know** what factorise means and create a better impression on the marker.)

9·7 the general features of the graphs of simple polynomial functions

Example

Sketch the general features of the curve with equation $f(x) = x(x-1)^2$

Solution there are zeros at $0, 1, 1,$
the double root indicates a tangent at $(1, 0)$.
$f\left(\frac{1}{2}\right) > 0 \Rightarrow$ there is a maximum turning point between 0 and 1.

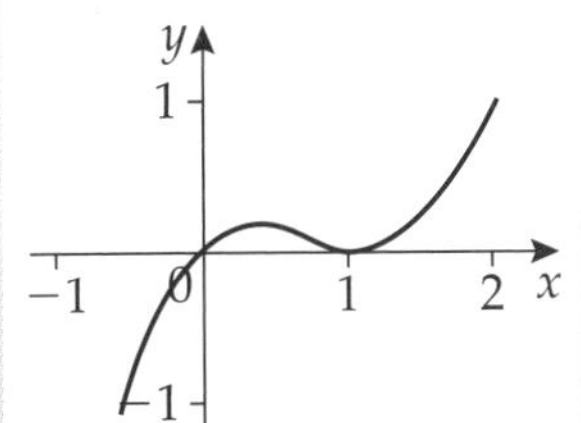

(Compare again with item **4·11**.)

9·8 how to prove that an equation has a root between two given (appropriate) values, and find that root to any required degree of accuracy

Example

Show that the equation $x^3 - x + 1 = 0$ has a root between -2 and -1 and find this root correct to 3 significant figures

(Solution) Let $x^3 - x + 1 = f(x)$. Then $f(-2) = -8 + 2 + 1 = -5\ (<0)$
and $f(-1) = -1 + 1 + 1 = 1\ (>0)$

$f(x)$ changes sign between -2 and -1, so there is a root between -2 and -1

(Note that this is only a valid conclusion because we know that the graph of a polynomial function is continuous.)
It is now convenient to use a programmable calculator.
Using the ANS key on a scientific calculator will take more time.

$f(-1{\cdot}3) = 0{\cdot}103\ (>0)$
$f(-1{\cdot}4) = -0{\cdot}344\ (<0) \Rightarrow$ root lies between $-1{\cdot}3$ and $-1{\cdot}4$

(Note that I tried several 2 digit numbers before writing down the two which involved a change of sign, namely $-1{\cdot}3$ and $-1{\cdot}4$.)

$f(-1{\cdot}32) = 0{\cdot}02\ldots\ (>0)$
$f(-1{\cdot}33) = -0{\cdot}0226\ldots(<0) \Rightarrow$ root lies between $-1{\cdot}32$ and $-1{\cdot}33$

(Similarly I did not need to write down all the 3 digit numbers I tried.)

(We now know that the root lies between these two 'consecutive' decimals with 3 significant figures. We decide which one the root is closer to by examining the value of $f(x)$ half way between them.

$f(-1{\cdot}3245) = 0{\cdot}001\ldots(>0) \Rightarrow$ root lies between $-1{\cdot}32$ and $-1{\cdot}325$
$\Rightarrow -1{\cdot}32$ correct to 3 significant figures

Chapter 10

INTEGRATION

What You Should Know

10·1 that integration is the inverse process of differentiation

10·2 that if $f(x) = F'(x)$ then $\int f(x)dx = F(x) + c$ where c is the constant of integration; this is an indefinite integral, and $\int_a^b f(x)\,dx = F(b) - F(a)$; this is a definite integral with limits a and b

10·3 how to determine the integrals of functions of the form px^n, $n \neq -1$, and the sum or difference of such functions.

10·4 how to solve equations of the form $\frac{dy}{dx} = f(x)$ [for suitable $f(x)$]

10·5 how to evaluate definite integrals

10·6 that the area bounded by the curve $y = f(x)$, the lines $x = a$, $x = b$ $(b > a)$ and the x-axis is $\int_a^b f(x)\,dx$ (for $f(x) \geqslant 0$ in $a \leqslant x \leqslant b$) and $-\int_a^b f(x)\,dx$ (for $f(x) \leqslant 0$ in $a \leqslant x \leqslant b$)

10·7 that if $f(x) \geqslant g(x)$ for $a \leqslant x \leqslant b$, then the area enclosed by the curves $y = f(x)$, $y = g(x)$ and the lines $x = a$, $x = b$ is $\int_a^b [f(x) - g(x)]dx$

10·8 how to apply the above in context

Notes on items 10·1 to 10·8: Integration

10·1 that integration is the inverse process of differentiation

Given that the derivative of $2x^3 + 6x^2 + 5x - 3$ is $6x^2 + 12x + 5$, the integral of $6x^2 + 12x + 5$ is $2x^3 + 6x^2 + 5x + c$, where c is the constant of integration. We cannot tell that the value of c is -3, because the derivative of $2x^3 + 6x^2 + 5x + 17$ for example is also $6x^2 + 12x + 5$.

10·2 that if $f(x) = F'(x)$ then $\int f(x)dx = F(x) + c$ where c is the constant of integration; this is an indefinite integral; and $\int_a^b f(x)\,dx = F(b) - F(a)$; this is a definite integral with limits a and b

Example

let $F(x) = 2x^3 + 6x^2 + 5x$, so $f(x) = F'(x) = 6x^2 + 12x + 5$

$\therefore \int f(x)dx = \int (6x^2 + 12x + 5)dx = F(x) + c = 2x^3 + 6x^2 + 5x + c,$

and

$$\int_1^2 f(x)dx = \int_1^2 (6x^2 + 12x + 5)dx = F(2) - F(1) = \left[2x^3 + 6x^2 + 5x\right]_1^2 = 50 - 13 = 37.$$

10·3 how to determine the integrals of functions of the form px^n, $n \neq -1$, and the sum or difference of such functions

Example

Integrate a) $(x+1)(x-3)$, b) $\dfrac{x^2 + 2x - 1}{\sqrt{x}}$ with respect to x.

(Solution) **Expressions like this have to be made into a string of terms, just as for differentiation.**

a) $\int (x+1)(x-3)dx = \int (x^2 - 2x - 3)dx = \dfrac{x^3}{3} - 2\left(\dfrac{x^2}{2}\right) - 3x + c = \dfrac{1}{3}x^3 - x^2 - 3x + c$

b) $\int \dfrac{x^2 + 2x - 1}{\sqrt{x}} dx = \int \dfrac{x^2 + 2x - 1}{x^{\frac{1}{2}}} dx = \int \left(\dfrac{x^2}{x^{\frac{1}{2}}} + \dfrac{2x}{x^{\frac{1}{2}}} - \dfrac{1}{x^{\frac{1}{2}}}\right) dx$

$= \int (x^{\frac{3}{2}} + 2x^{\frac{1}{2}} - x^{-\frac{1}{2}})dx$

$= \dfrac{x^{\frac{5}{2}}}{\frac{5}{2}} + 2\left(\dfrac{x^{\frac{3}{2}}}{\frac{3}{2}}\right) - \dfrac{x^{\frac{1}{2}}}{\frac{1}{2}} + c = \dfrac{2}{5}x^{\frac{5}{2}} + \dfrac{4}{3}x^{\frac{3}{2}} - 2x^{\frac{1}{2}} + c$

10·4 how to solve equations of the form $\dfrac{dy}{dx} = f(x)$ [for suitable $f(x)$]

Example

Solve $\dfrac{dy}{dx} = 6x^2 + 4x$ and find the particular solution for which $y = 6$ when $x = 1$.

(Solution) $y = \int (6x^2 + 4x)dx = 2x^3 + 2x^2 + c$

$x = 1$ when $y = 6 \Rightarrow 6 = 2 + 2 + c \Rightarrow c = 2$

$\Rightarrow y = 2x^3 + 2x^2 + 2$

10·5 how to evaluate definite integrals

Example

Evaluate $\int_1^4 \frac{4}{x\sqrt{x}}\,dx$.

(Solution) $\int_1^4 \frac{4}{x\sqrt{x}}\,dx = \int_1^4 \frac{4}{x^{\frac{3}{2}}}\,dx = \int_1^4 4x^{-\frac{3}{2}}\,dx = \left[4\left(\frac{x^{-\frac{1}{2}}}{-\frac{1}{2}}\right)\right]_1^4 = -8\left[\frac{1}{\sqrt{x}}\right]_1^4 = -8\left[\frac{1}{2} - 1\right] = 4$

10·6 that the area bounded by the curve $y = f(x)$, the lines $x = a$, $x = b$ ($b > a$) and the x-axis is $\int_a^b f(x)dx$ (for $f(x) \geqslant 0$ in $a \leqslant x \leqslant b$) and $-\int_a^b f(x)dx$ (for $f(x) \leqslant 0$ in $a \leqslant x \leqslant b$)

Example

Calculate the area of the shaded region shown.

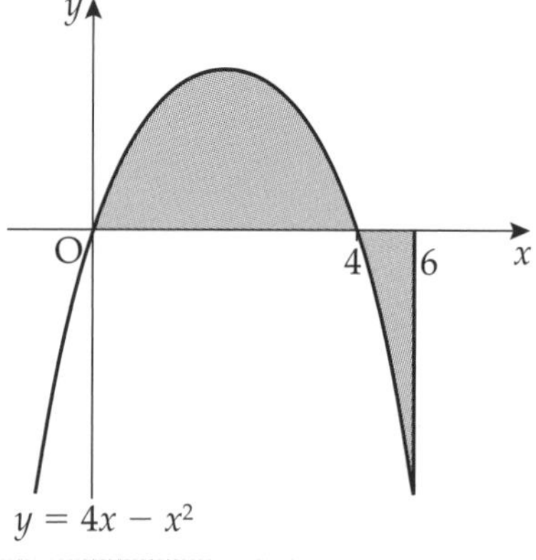

(Solution) The areas above and below the x-axis must be calculated separately.

$$\text{upper area} = \int_0^4 (4x - x^2)dx$$
$$= \left[2x^2 - \frac{x^3}{3}\right]_0^4$$
$$= \left[32 - \frac{64}{3}\right] - 0 = \frac{32}{3}$$

$$\text{lower area} = -\int_4^6 (4x - x^2)dx$$
$$= \left[-2x^2 + \frac{x^3}{3}\right]_4^6$$
$$= [-72 + 72] - \left[-32 + \frac{64}{3}\right]$$
$$= 32 - \frac{64}{3} = \frac{32}{3}$$

[I prefer to make use of $\int_a^b f(x)dx = -\int_b^a f(x)dx$ and write

$$-\int_4^6 (4x - x^2)dx = \left[2x^2 - \frac{x^3}{3}\right]_6^4 = \left[32 - \frac{64}{3}\right] - [72 - 72] = \frac{32}{3}]$$

hence total area $= \frac{32}{3} + \frac{32}{3} = \frac{64}{3} = 21\frac{1}{3}$ units^2

10·7 **that if $f(x) \geqslant g(x)$ for $a \leqslant x \leqslant b$, then the area enclosed by the curves $y = f(x)$, $y = g(x)$ and the lines $x = a$, $x = b$ is $\int_a^b [f(x) - g(x)]dx$**

Example

Calculate the shaded area shown.

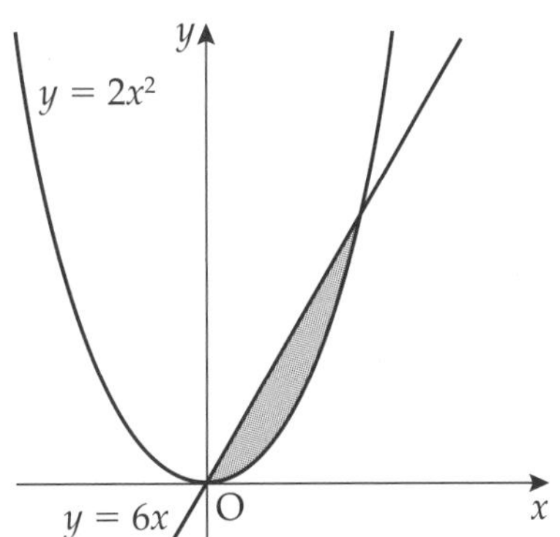

(Solution) First we have to find the x-coords of the points of intersection.

$$2x^2 = 6x$$
$$\Rightarrow 2x(x - 3) = 0$$
$$\Rightarrow x = 0, 3$$

these are the limits for the integration

$$\text{Area} = \int_0^3 (6x - 2x^2)dx$$

It is important to integrate (the upper curve – the lower curve). Basically you are subtracting the area under the lower curve from the area under the upper curve. This is worth a mark in the exam. The answer should always be positive. If you obtain a negative answer, go back and check for an error. If you don't pay any attention to the first line, you do not know if a negative answer indicates an error or not.

$$\text{Thus area} = \int_0^3 (6x - 2x^2)dx = \left[3x^2 - \frac{2x^3}{3}\right]_0^3 = [27 - 18] - 0 = 9 \text{ units}^2$$

10·8 **how to apply the above in context.**

See Chapter 18 on typical longer exam questions: (Question **18·4**).

Remember

Learning mathematical skills is an activity in which you actually have to become involved yourself before it registers in your mind.

You **learn** mathematics when you do it for yourself and when you discuss it with others.

Remember that getting stuck is part of learning. It is how you get unstuck that helps you to understand and remember things.

Chapter 11

COMPOUND AND MULTIPLE ANGLES

What You Should Know

11·1 the multiple angle formulae

$$\begin{aligned}
\sin(A+B) &= \sin(A)\cos(B) + \cos(A)\sin(B)\\
\sin(A-B) &= \sin(A)\cos(B) - \cos(A)\sin(B)\\
\cos(A+B) &= \cos(A)\cos(B) - \sin(A)\sin(B)\\
\cos(A-B) &= \cos(A)\cos(B) + \sin(A)\sin(B)
\end{aligned}$$

$$\begin{aligned}
\sin(2A) &= 2\sin(A)\cos(A)\\
\cos(2A) &= \cos^2(A) - \sin^2(A)\\
&= 2\cos^2(A) - 1\\
&= 1 - 2\sin^2(A)\\
\cos^2(A) &= \frac{1}{2}[1 + \cos(2A)]\\
\sin^2(A) &= \frac{1}{2}[1 - \cos(2A)]
\end{aligned}$$

11·2 how to apply these formulae in numerical cases

eg. given $\tan(x) = \frac{4}{3}$, $\tan(y) = \frac{5}{12}$, find the exact value of $\sin(x-y)$ and $\sin(2x)$

11·3 how to apply these formulae to solve equations

eg. $\sin(2x) = \cos(x)$, $\cos(2x) - \sin(x) + 1 = 0$

11·4 how to apply these formulae in the solution of geometrical problems

Notes on items 11·1 to 11·4: Compound and Multiple Angles

11·1 the multiple angle formulae

Key Points

$$\begin{aligned}
\sin(A+B) &= \sin(A)\cos(B) + \cos(A)\sin(B)\\
\sin(A-B) &= \sin(A)\cos(B) - \cos(A)\sin(B)\\
\cos(A+B) &= \cos(A)\cos(B) - \sin(A)\sin(B)\\
\cos(A-B) &= \cos(A)\cos(B) + \sin(A)\sin(B)
\end{aligned}$$

(Also)

$$\begin{aligned}
\sin(2A) &= 2\sin(A)\cos(A)\\
\cos(2A) &= \cos^2(A) - \sin^2(A)\\
&= 2\cos^2(A) - 1\\
&= 1 - 2\sin^2(A)
\end{aligned}$$

(and so)

$$\cos^2(A) = \frac{1}{2}[1 + \cos(2A)]$$

$$\sin^2(A) = \frac{1}{2}[1 - \cos(2A)]$$

These formulae must be well known, even though most of them appear in the formula list on the exam paper. Remember that you have to recognise the right hand side as well. I do not recommend relying on the formula list. They are important; learn them.

Think of the $\sin(2A)$ formula as $\sin(\text{angle}) = 2 \times \sin(\text{half the angle}) \times \cos(\text{half the angle})$.

The final two formulae are deduced by changing the subject of the two previous formulae. They allow a quadratic function (in $\sin x$ or $\cos x$) to be expressed as a linear trig function (in $\cos 2x$). This process is important for integration for example.

11·2 how to apply these formulae in numerical cases

Example

Given $\tan(x) = \frac{4}{3}$ and $\tan(y) = \frac{5}{12}$, where x and y are acute angles, find the exact value of $\sin(x - y)$ and $\sin(2x)$.

(Solution) If we wish to use the formula for $\sin(x - y)$, we need to know the values of $\sin(x)$, $\cos(x)$, $\sin(y)$, $\cos(y)$.

Since x and y are acute, the usual strategy is to draw right angled triangles containing angles equal to x and y; use Pythagoras' Theorem to find the third side and hence all the trig ratios we require.

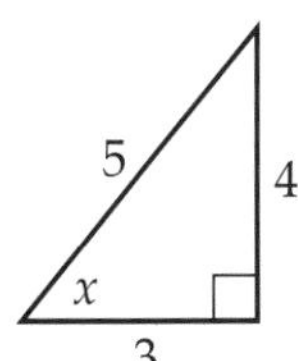

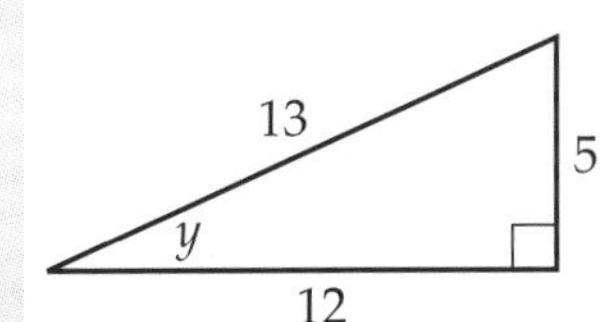

$$\sin(x - y) = \sin(x)\cos(y) - \cos(x)\sin(y) = \frac{4}{5} \times \frac{12}{13} - \frac{3}{5} \times \frac{5}{13} = \frac{48 - 15}{65} = \frac{33}{65}$$

$$\sin(2x) = 2\sin x \cos x = 2\left(\frac{4}{5}\right)\left(\frac{3}{5}\right) = \frac{24}{25}$$

11·3 how to apply these formulae to solve equations

Example

Solve

a) $\sin(2x) = \cos(x)$ *for* $0 \leq x \leq 2\pi$ (This could be in Paper 1)

b) $\cos(2x°) - \sin(x°) + 1 = 0$ for $0 \leq x \leq 360$.

***Example** continued ➢*

***Example** continued*

(Solution) a) $\sin(2x) = \cos(x)$

$\Rightarrow 2\sin(x)\cos(x) = \cos(x)$ – (use the $\sin(2x)$ formula)

$\Rightarrow 2\sin(x)\cos(x) - \cos(x) = 0$ – (take everything to one side)*

$\Rightarrow \cos(x)[2\sin(x) - 1] = 0$ – (take out the common factor)

$\Rightarrow \cos(x) = 0 \text{ or } \sin(x) = \frac{1}{2}$

$\Rightarrow x = \frac{\pi}{2}, \frac{3\pi}{2} \text{ or } x = \frac{\pi}{6}, \frac{5\pi}{6}$ i.e. $x = \frac{\pi}{6}, \frac{\pi}{2}, \frac{5\pi}{6}, \frac{3\pi}{2}$

* instead of this you can say 'either $\cos(x) = 0$ or $2\sin(x) = 1$' remembering that you can only divide both sides of an equation by a non-zero quantity. If you omit the '$\cos(x) = 0$' you lose half the solutions (and possibly half the marks).

b) $\cos(2x°) - \sin(x°) + 1 = 0$

$\Rightarrow [1 - 2\sin^2(x°)] - \sin(x°) + 1 = 0$ – (choose the formula with $\sin(x)$)

$\Rightarrow 2\sin^2(x°) + \sin(x°) - 2 = 0$ – (arrange in standard form)

$$\Rightarrow \sin(x°) = \frac{-1 \pm \sqrt{1 - 4 \times 2(-2)}}{4} = \frac{-1 \pm \sqrt{17}}{4}$$

$$\Rightarrow \sin(x°) = \frac{-1 + \sqrt{17}}{4} \text{ or } \frac{-1 - \sqrt{17}}{4}$$

$\Rightarrow 0{\cdot}780\ldots \text{ or } -1{\cdot}280$

$\Rightarrow x = 51{\cdot}3, 128{\cdot}7$ no solutions to $\sin x = -1{\cdot}280$

11·4 how to apply these formulae in the solution of geometrical problems.

See Chapter 18 on typical longer exam questions: (Question 18·5).

Chapter 12

THE CIRCLE

What You Should Know

12·1 that the equation of the circle with centre the origin and radius r is $x^2 + y^2 = r^2$

12·2 that the equation of the circle with centre (a, b) and radius r is $(x - a)^2 + (y - b)^2 = r^2$

12·3 that the equation $x^2 + y^2 + 2gx + 2fy + c = 0$ represents the circle with centre $(-g, -f)$, radius $\sqrt{g^2 + f^2 - c}$ (provided $g^2 + f^2 > c$)

12·4 how to find the equation of a circle given the centre and radius, or equivalent information e.g. the centre and a point on the circumference, the ends of a diameter, 3 points on the circumference, etc.

12·5 how to find the equation of a tangent at a given point on the circumference

12·6 how to determine whether two given circles touch each other

12·7 how to determine the points of intersection of a line and a circle

12·8 how to determine whether or not a line is a tangent to a circle

12·9 how to apply the above in context

Notes on items 12·1 to 12·9: The Circle

12·1 that the equation of the circle with centre the origin and radius r is $x^2 + y^2 = r^2$

The equation of the circle with centre the origin and radius 5 is $x^2 + y^2 = 25$.

You must also be able to recognise $x^2 + y^2 = 49$ as the equation of the circle with centre the origin and radius 7.

12·2 that the equation of the circle with centre (a, b) and radius r is $(x - a)^2 + (y - b)^2 = r^2$

The equation of the circle with centre $(4, -3)$ and radius 8 is $(x - 4)^2 + (y + 3)^2 = 64$.

You must also recognise $(x + 7)^2 + (y - 2)^2 = 16$ as the equation of the circle with centre $(-7, 2)$ and radius 4.

The formula in item 12·1 is a special case of $(x-a)^2+(y-b)^2=r^2$ with a and b equal to 0.

This is the formula that you should always use when finding the equation of a circle.

12·3 that the equation $x^2+y^2+2gx+2fy+c=0$ represents the circle with centre $(-g,-f)$ and radius $\sqrt{g^2+f^2-c}$ (provided $g^2+f^2>c$)

Example

This is known as the general equation of the circle.

For the equation $x^2+y^2+2x-8y+8=0$, 'g' $=1$, 'f' $=-4$, 'c' $=8$.

So this represents the circle with centre $(-1, 4)$ and radius $\sqrt{1^2+(-4)^2-8}=\sqrt{9}=3$.

Notice that $x^2+y^2+2x-8y+18=0$ does **not** represent a circle because $g^2+f^2<c$, i.e. $r=\sqrt{1+16-18}=\sqrt{-1}$ which is not a real number

Note also: for $3x^2+3y^2+12x-8y+7=0$, divide throughout by 3 first of all.

12·4 how to find the equation of a circle given the centre and radius, or equivalent information

Example

Find the equation of the circle

a) with centre the origin and passing through the point $(4, 5)$
b) with centre $(4, 5)$ and passing through the origin
c) where P $(2, 7)$ and Q $(8, -1)$ are opposite ends of a diameter
d) passing through the points $(0, 2)$, $(0, 8)$ and touching the positive x-axis.

(Solution) a)

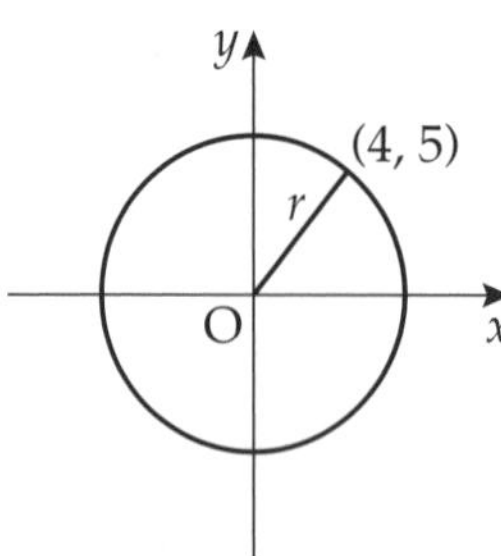

$r^2=4^2+5^2=41$
$\Rightarrow x^2+y^2=41$

b)

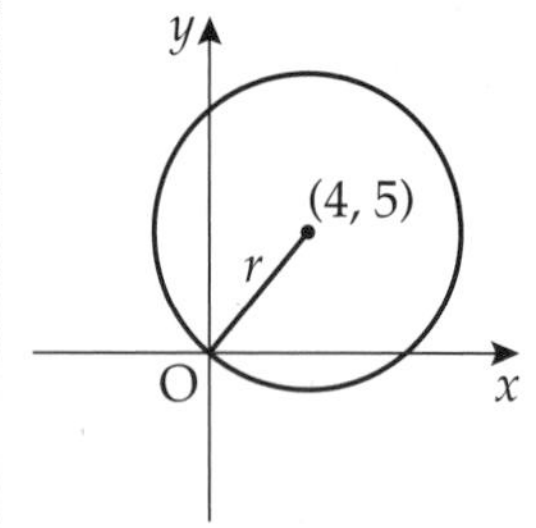

$r^2=4^2+5^2=41$
$\Rightarrow (x-4)^2+(y-5)^2=41$

Example continued ➢

Example *continued*

c)

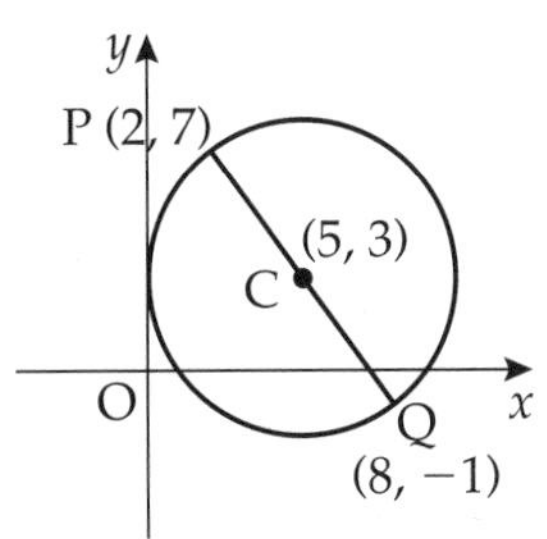

the centre, C, is the mid point of PQ i.e. (5, 3)
the radius = CP = $\sqrt{9+16} = 5$
$\Rightarrow (x-5)^2 + (y-3)^2 = 25$

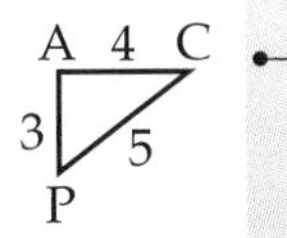

d) (the sketch is essential this time)

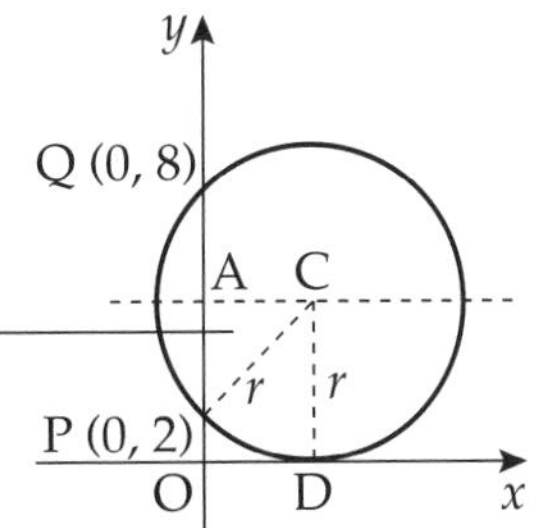

Since AC meets the chord PQ at 90°, it bisects it. Hence
A is (0, 5) and so C must lie on the line $y = 5$
so the radius CD must be 5;
so the radius PC is also 5;
AP = 5 − 2 = 3
thus we have a '3, 4, 5' triangle, and so the centre must be (4, 5),
and the equation $(x-4)^2 + (y-5)^2 = 25$

12·5 how to find the equation of a tangent at a given point on the circumference

Example Find the equation of the tangent to the circle with equation $x^2 + y^2 - 10x - 6y + 5 = 0$ at the point (3, −2).

(Solution) Just because the word tangent appears, do not jump to the conclusion that differentiation is required. To obtain the equation of a line we need a point on the line and its gradient.

P (3, −2) is the point, and the tangent is perpendicular to the radius.
To find the gradient of the radius, we need the coordinates of the centre.
Make a sketch of this, possibly even without the circle itself.

Let C be the centre of the circle, Then C is (5, 3)

$$m_{CP} = \frac{3-(-2)}{5-3} = \frac{5}{2}$$

$$\Rightarrow m_{tgt} = \frac{-1}{m_{CP}} = -\frac{2}{5}$$

$\Rightarrow$ equation of tangent is $y - (-2) = -\frac{2}{5}(x-3)$
(remember: clear the fractions first:)

$$5y + 10 = -2x + 6$$

$$\Rightarrow 2x + 5y + 4 = 0$$

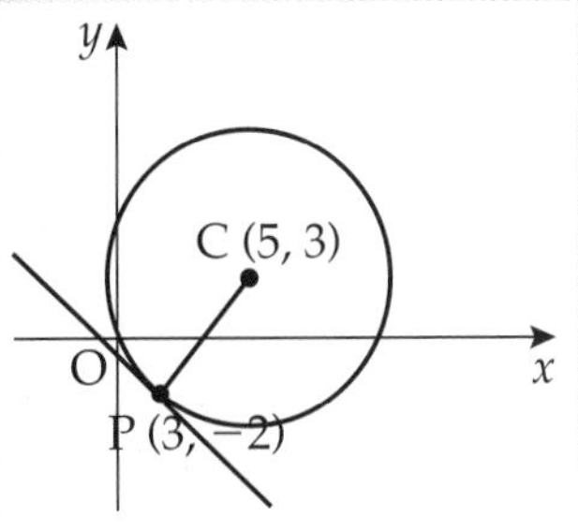

12·6 how to determine whether two given circles touch each other

If two circles touch externally, then the distance between their centres equals the sum of their radii; i.e. $AB = r_1 + r_2$.

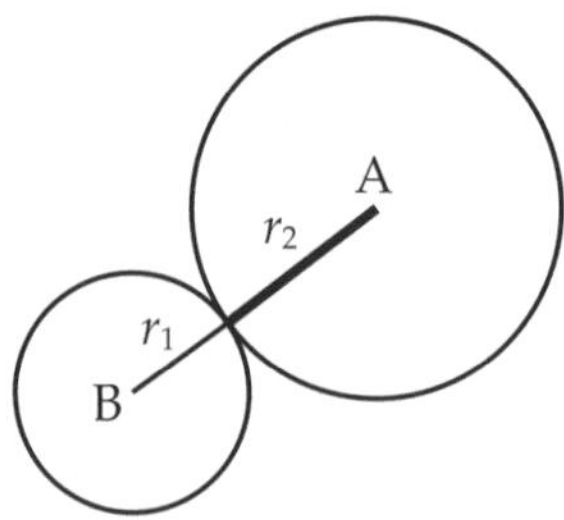

If two circles touch, one inside the other, then the distance between their centres equals the difference between their radii; i.e. $AB = r_1 - r_2$.

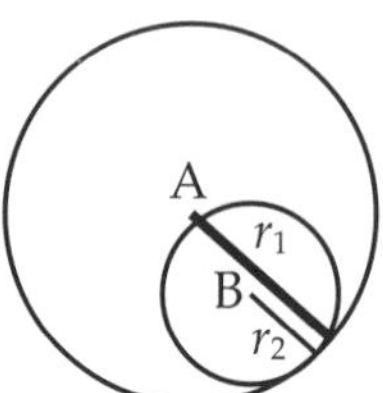

Example

Show that the circles C_1 and C_2 with equations
C_1: $(x+3)^2 + (y+2)^2 = 20$ and C_2: $(x-11)^2 + (y-5)^2 = 5$
touch the circle C_3 with equation $(x-7)^2 + (y-3)^2 = 45$.

(Solution) Let the circles C_1, C_2, C_3 have centres A, B, C and radii r_1, r_2, r_3 respectively.

A is $(-3, -2)$, and $r_1 = \sqrt{20} = 2\sqrt{5}$, B $(11, 5)$, and $r_2 = \sqrt{5}$, C $(7, 3)$ and $r_3 = \sqrt{45} = 3\sqrt{5}$.

$AC = \sqrt{(-3-7)^2 + (-2-3)^2} = \sqrt{125} = 5\sqrt{5}$;

$r_1 + r_3 = 2\sqrt{5} + 3\sqrt{5} = 5\sqrt{5} = AC$; hence C_1 and C_3 touch externally

$BC = \sqrt{(11-7)^2 + (5-3)^2} = \sqrt{20} = 2\sqrt{5}$

$r_3 - r_2 = 3\sqrt{5} - \sqrt{5} = 2\sqrt{5} = BC$ hence C_3 and C_2 touch with C_2 inside C_3.

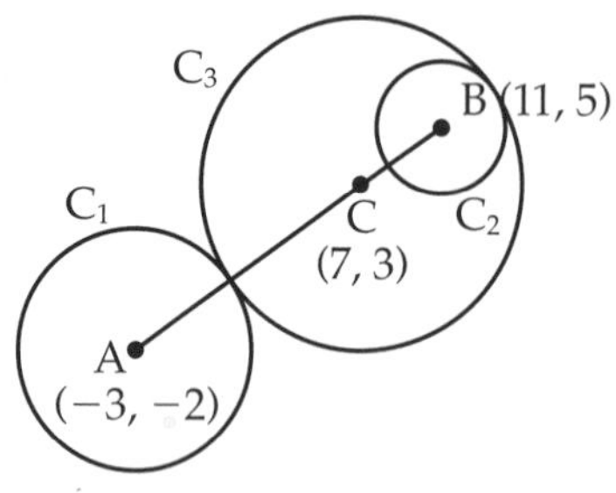

12·7 how to determine the points of intersection of a line and a circle

The method is exactly the same as for determining the intersection of a line and a parabola (see item 8·7), except that with a circle equation there could be a y term as well as the y^2 term, so you must substitute for both of them.

Example Find the points of intersection of the line with equation $3x - y = 13$ and the circle with equation $x^2 + y^2 - 4x - 6y - 7 = 0$.

(Solution) (Carefully re-arrange the linear equation to make either x or y the subject. In this case choose y in order to avoid fractions.)

$3x - y = 13 \Rightarrow 3x = 13 + y \Rightarrow 3x - 13 = y \Rightarrow y = 3x - 13$

substitute for each occurrence of y in the circle equation

i.e. $x^2 + (3x - 13)^2 - 4x - 6(3x - 13) - 7 = 0$

$\Rightarrow 10x^2 - 100x + 240 = 0$

$\Rightarrow x^2 - 10x + 24 = 0$ *(I will refer to this line in item 12·8.)

$\Rightarrow (x - 4)(x - 6) = 0 \Rightarrow x = 4, 6$

$\Rightarrow y = -1, 5$ hence $(4, -1)$ and $(6, 5)$

12·8 how to determine whether or not a line is a tangent to a circle

(In item 12·7 we saw an example of a line cutting a circle in two points. Notice that at the equation marked *, the value of the discriminant is $100 - 4 \times 1 \times 24 = 4$ (>0) which indicates distinct real roots. This is consistent with there being two points of intersection. A different possibility is that the discriminant is negative, giving no real roots and hence the line does not intersect the circle at all. The third and final possibility is that the roots are equal, which means that the two points of intersection coincide, so the line is a tangent to the circle.)

Example Show that the line with equation $2x + y = 17$ is a tangent to the circle with equation $x^2 + y^2 - 4x - 6y - 7 = 0$.

(Solution) As above, carefully re-arrange the linear equation: $y = 17 - 2x$;

substitute: $x^2 + (17 - 2x)^2 - 4x - 6(17 - 2x) - 7 = 0$

$\Rightarrow x^2 + 289 - 68x + 4x^2 - 4x - 102 + 12x - 7 = 0$

$\Rightarrow 5(x^2 - 12x + 36) = 0$

$\Rightarrow \Delta = 144 - 4 \times 1 \times 36 = 0 \Rightarrow$ equal roots

$\Rightarrow$ the line is a tangent to the circle

(If you wish to find the point of contact, solve the equation

i.e. $(x - 6)^2 = 0 \Rightarrow x = 6, 6 \Rightarrow y = 5 \Rightarrow (6, 5)$

You can omit reference to Δ and justify the tangency with this working if you communicate by stating 'equal roots' after the $x = 6, 6$.)

[Refer again to item 8·7.]

12·9 how to apply the above in context

(See Chapter 18 on typical longer exam questions:
Examples 18·6, 18·7, 18·17.)

Remember *Show all your working.*

How many times have you heard that now? Don't lose marks for not convincing your markers that you thoroughly understand what you are doing.

Chapter 13

VECTORS

What You Should Know

13·1 the meaning of the terms vector, scalar, position vector, unit vector, components

13·2 how to add and subtract vectors, and find a scalar multiple of a vector

13·3 that vectors $\mathbf{u}$ and $\mathbf{v}$ are parallel $\Leftrightarrow \mathbf{v} = k\mathbf{u}$ for some scalar k

13·4 how to calculate the components of $\overrightarrow{AB}$ given the coordinates of A and B

13·5 that $\mathbf{i}$, $\mathbf{j}$, $\mathbf{k}$ are the unit vectors parallel to the x, y, z-axis respectively and that the column vector $\begin{pmatrix} a \\ b \\ c \end{pmatrix}$ is equivalent to $a\mathbf{i} + b\mathbf{j} + c\mathbf{k}$

13·6 that $\begin{pmatrix} a \\ b \\ c \end{pmatrix} = \begin{pmatrix} d \\ e \\ f \end{pmatrix} \Leftrightarrow a = d, b = e, c = f$

13·7 how to calculate the length of a vector, or the distance between two points in space

13·8 how to determine whether or not three points in space are collinear, and, if collinear, to determine the ratio in which one divides the join of the other two

13·9 how to calculate the co-ordinates of a point which divides the line joining two given points in a given ratio (preferably by using the section formula)

13·10 that the scalar product is defined by $\mathbf{a}.\mathbf{b} = |\mathbf{a}|\,|\mathbf{b}|\cos\theta$ where θ is the angle between the positive directions of $\mathbf{a}$ and $\mathbf{b}$

13·11 that $\mathbf{a}.\mathbf{b} = a_1b_1 + a_2b_2 + a_3b_3$ also defines the scalar product

13·12 how to calculate the angle between two vectors (given in component form)

13·13 that the scalar product is distributive over addition, i.e. $\mathbf{a}.(\mathbf{b} + \mathbf{c}) = \mathbf{a}.\mathbf{b} + \mathbf{a}.\mathbf{c}$

13·14 that for $\mathbf{a}, \mathbf{b} \neq 0$, $\mathbf{a}.\mathbf{b} = 0 \Leftrightarrow \mathbf{a}$ is perpendicular to $\mathbf{b}$

13·15 how to use vectors to justify geometrical results and apply vectors in problem situations

Notes on items 13·1 to 13·15: Vectors

13·1 the meaning of the terms vector, scalar, position vector, unit vector, components

Key Words and *Definitions*

A **vector** quantity has magnitude, direction and sense. The sense can be specified within the direction, e.g. eastwards. Examples of vector quantities are velocity, acceleration, force. The notations used to denote vectors are $\overrightarrow{AB}$ or **u** or $\underline{u}$ or $\underset{\sim}{u}$. Pupils are notoriously bad at making proper use of notations, which sometimes cause them confusion between numbers and vectors. It is not helped by the fact that most text books use the bold type notation, which cannot be replicated in manuscript. Perhaps it would be better if school texts on vectors stuck to the underlining notation.

The magnitude of a vector is a **scalar** quantity. Examples of scalar quantities include time, mass, work done. A scalar possesses only size, no direction.

The **position vector** of a point in space is the vector from the origin to the point. It is usually given the same name as the point but in lower case, e.g. the point C has position vector **c**. A **unit vector** is any vector whose length is 1 unit. A vector can be represented by a directed line segment. The magnitude direction and sense of a vector can all be conveyed by **components**, thus:

in a plane

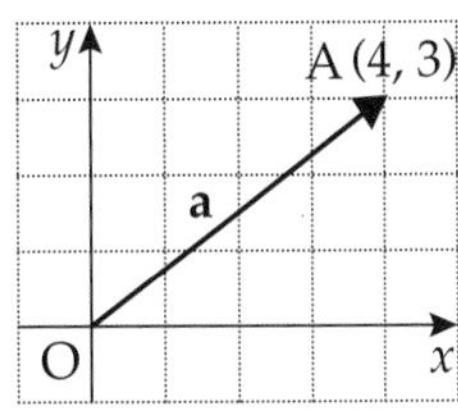

For the point A (4, 3), $\overrightarrow{OA} = \mathbf{a} = \begin{pmatrix} 4 \\ 3 \end{pmatrix}$

a is the position vector of A

in space

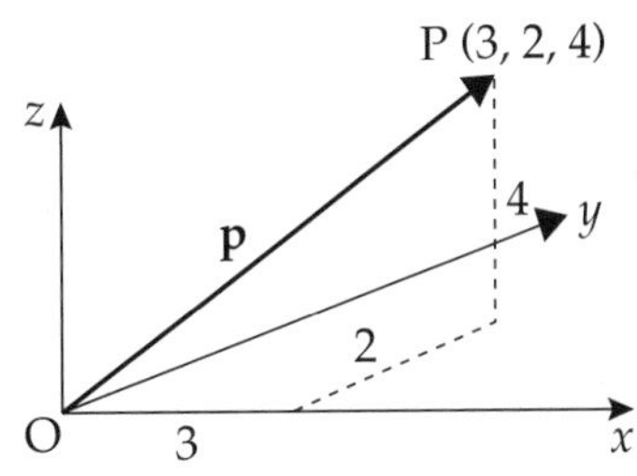

For the point P (3, 2, 4), $\overrightarrow{OP} = \mathbf{p} = \begin{pmatrix} 3 \\ 2 \\ 4 \end{pmatrix}$

p is the position vector of P

Vectors which are not being used as position vectors are called free vectors. Vectors which have the same components are considered to be equal, for example **a** above would be equal to $\overrightarrow{ST}$ where S is (1, 1) and T (5, 4). The directed line segments $\overrightarrow{OA}$ and $\overrightarrow{ST}$ are two of the many representatives of **a**.

The length of OA is clearly 5, so the unit vector in the OA direction is $\frac{1}{5}\overrightarrow{OA}$ or $\begin{pmatrix} \frac{4}{5} \\ \frac{3}{5} \end{pmatrix}$

Remember

- use either a single lower case letter (a small letter) or two upper case letters (capital letters) for your vectors. $\overrightarrow{A}$ does not make sense.
- underline your vectors. Clarity in this will help you in later examples.
- write co-ordinates horizontally and components vertically (as above).

13·2 how to add and subtract vectors, and find a scalar multiple of a vector

We **add** vectors by putting them nose to tail.
We can remember this as either the triangle law or the parallelogram law for the addition of vectors. In practice, it is more usual just to add the corresponding components,

e.g. if $\mathbf{u} = \begin{pmatrix} 3 \\ 4 \end{pmatrix}$ and $\mathbf{v} = \begin{pmatrix} 2 \\ -1 \end{pmatrix}$, then $\mathbf{u} + \mathbf{v} = \begin{pmatrix} 3 \\ 4 \end{pmatrix} + \begin{pmatrix} 2 \\ -1 \end{pmatrix} = \begin{pmatrix} 5 \\ 3 \end{pmatrix}$

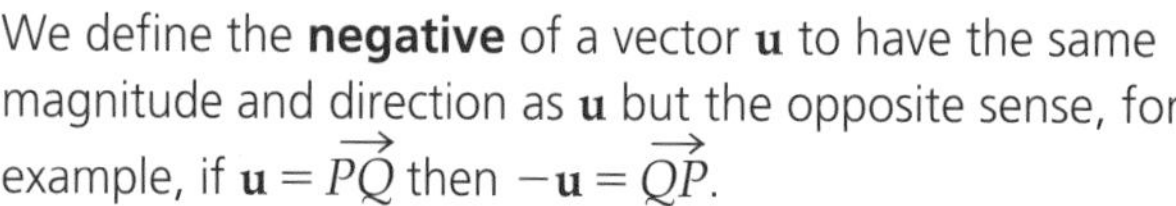

We define the **negative** of a vector $\mathbf{u}$ to have the same magnitude and direction as $\mathbf{u}$ but the opposite sense, for example, if $\mathbf{u} = \overrightarrow{PQ}$ then $-\mathbf{u} = \overrightarrow{QP}$.

In practice, it is more usual just to change the signs of the components,

e.g. $\mathbf{u} = \begin{pmatrix} 7 \\ -2 \end{pmatrix} \Rightarrow -\mathbf{u} = -\begin{pmatrix} 7 \\ -2 \end{pmatrix} = \begin{pmatrix} -7 \\ 2 \end{pmatrix}$

We define the **subtraction** of vectors thus:

$\boldsymbol{a} - \boldsymbol{b} = \boldsymbol{a} + (-\boldsymbol{b})$

In practice, it is more usual just to subtract the corresponding components,

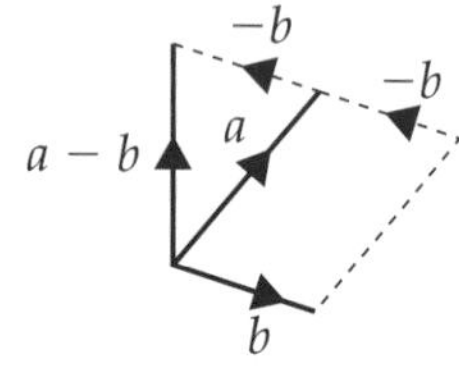

e.g. $\mathbf{u} = \begin{pmatrix} 7 \\ 11 \\ 4 \end{pmatrix}, \mathbf{v} = \begin{pmatrix} 3 \\ -1 \\ 2 \end{pmatrix} \Rightarrow \mathbf{u} - \mathbf{v} = \begin{pmatrix} 7 \\ 11 \\ 4 \end{pmatrix} - \begin{pmatrix} 3 \\ -1 \\ 2 \end{pmatrix} = \begin{pmatrix} 4 \\ 12 \\ 2 \end{pmatrix}$

If we add k vectors each of which is $\mathbf{u}$, we get $k\mathbf{u}$.

Since k is a number and $\mathbf{u}$ is a vector, we call this **scalar multiplication** of a vector.

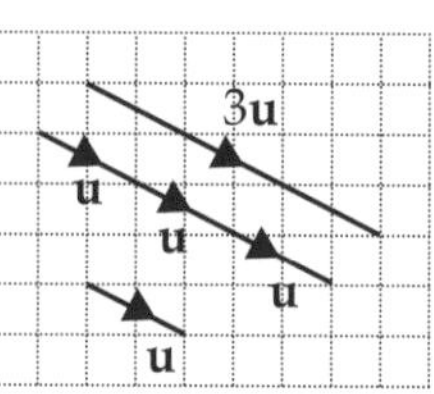

In practice, it is more usual just to multiply the components of $\mathbf{u}$ by k

e.g. $\mathbf{u} = \begin{pmatrix} 2 \\ -1 \end{pmatrix} \Rightarrow 3\mathbf{u} = 3\begin{pmatrix} 2 \\ -1 \end{pmatrix} = \begin{pmatrix} 6 \\ -3 \end{pmatrix}$

as shown in the diagram.

13·3 **that vectors u and v are parallel $\Leftrightarrow$ $\mathbf{v} = k\mathbf{u}$ for some scalar k**

See example **13·6**, which combines the concepts of items **13·3** and **13·6**.

13·4 **how to calculate the components of $\overrightarrow{AB}$ given the coordinates of A and B.**

$\overrightarrow{AB} = \overrightarrow{AO} + \overrightarrow{OB} = -\overrightarrow{OA} + \overrightarrow{OB} = -\mathbf{a} + \mathbf{b} = \mathbf{b} - \mathbf{a}$

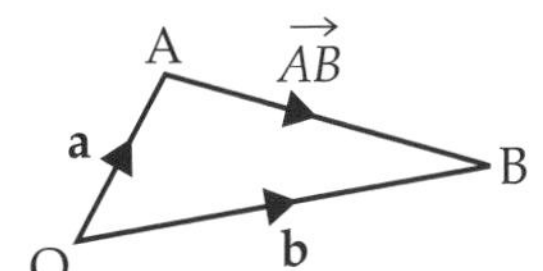

e.g. A (2, 3) and B (3, −4) $\Rightarrow \overrightarrow{AB} = \mathbf{b} - \mathbf{a} = \begin{pmatrix} 3 \\ -4 \end{pmatrix} - \begin{pmatrix} 2 \\ 3 \end{pmatrix} = \begin{pmatrix} 1 \\ -7 \end{pmatrix}$

13·5 **that i, j, k are the unit vectors parallel to the x, y, z-axes respectively and that the column vector $\begin{pmatrix} a \\ b \\ c \end{pmatrix}$ is equivalent to $a\mathbf{i} + b\mathbf{j} + c\mathbf{k}$**

In components, $\mathbf{i} = \begin{pmatrix} 1 \\ 0 \\ 0 \end{pmatrix}$, $\mathbf{j} = \begin{pmatrix} 0 \\ 1 \\ 0 \end{pmatrix}$, $\mathbf{k} = \begin{pmatrix} 0 \\ 0 \\ 1 \end{pmatrix}$

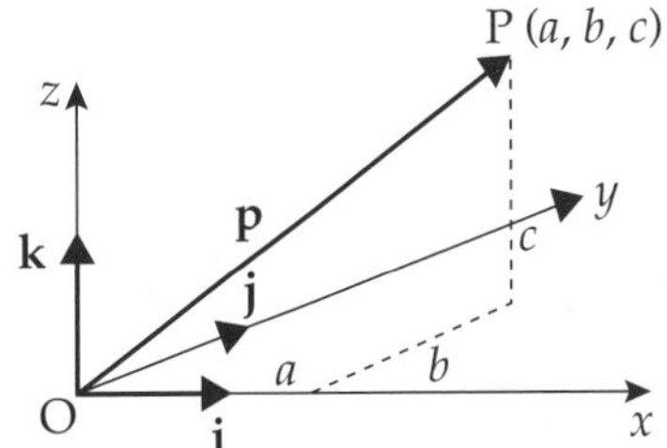

13·6 **that $\begin{pmatrix} a \\ b \\ c \end{pmatrix} = \begin{pmatrix} d \\ e \\ f \end{pmatrix} \Leftrightarrow a = d, b = e, c = f$**

Example A is the point (2, 3, −1), B (7, 4, 5), C (−4, 3, −5) and D (6, 5, z). Find the value of z for which AB is parallel to CD.

(Solution) $\overrightarrow{AB} = \mathbf{b} - \mathbf{a} = \begin{pmatrix} 7 \\ 4 \\ 5 \end{pmatrix} - \begin{pmatrix} 2 \\ 3 \\ -1 \end{pmatrix} = \begin{pmatrix} 5 \\ 1 \\ 6 \end{pmatrix}$ $\quad \overrightarrow{CD} = \mathbf{d} - \mathbf{c} = \begin{pmatrix} 6 \\ 5 \\ z \end{pmatrix} - \begin{pmatrix} -4 \\ 3 \\ -5 \end{pmatrix} = \begin{pmatrix} 10 \\ 2 \\ z+5 \end{pmatrix}$

$$AB \parallel CD \Rightarrow \overrightarrow{AB} \Rightarrow k\overrightarrow{CD} \Rightarrow \begin{pmatrix} 5 \\ 1 \\ 6 \end{pmatrix} = k\begin{pmatrix} 10 \\ 2 \\ z+5 \end{pmatrix} \Rightarrow \begin{aligned} 5 &= 10k \\ 1 &= 2k \\ 6 &= k(z+5) \end{aligned}$$

both of the first two equations give $\quad k = \frac{1}{2}$

substituting in the third $\Rightarrow 6 = \frac{1}{2}(z + 5) \Rightarrow z = 7$

13·7 how to calculate the length of a vector, or the distance between two points in space

The length of a vector

$$\mathbf{p} = \begin{pmatrix} a \\ b \\ c \end{pmatrix} \Rightarrow |\mathbf{p}| = \sqrt{OC^2 + CP^2}$$

$$= \sqrt{OD^2 + DC^2 + CP^2}$$

$$= \sqrt{a^2 + b^2 + c^2}$$

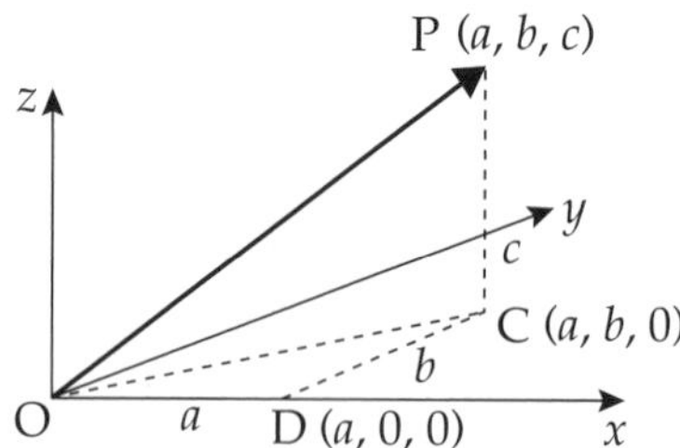

The distance formula

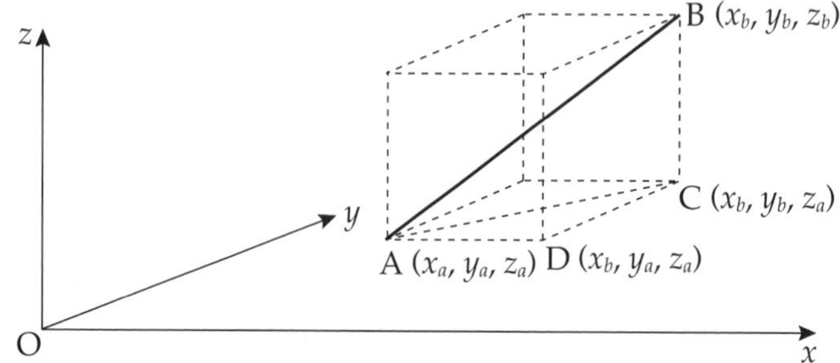

$$AB^2 = AC^2 + CB^2 = (AD^2 + DC^2) + CB^2 = (x_a - x_b)^2 + (y_a - y_b)^2 + (z_a - z_b)^2$$

$$\Rightarrow AB = \sqrt{(x_a - x_b)^2 + (y_a - y_b)^2 + (z_a - z_b)^2}$$

Note: if the z-co-ordinates are zero, this is the same formula as in item **3·1**.

Example

Find a) the length of the vector $\mathbf{u} = 3\mathbf{i} + 4\mathbf{j} - 7\mathbf{k}$

b) the distance between the points R (3, −4, 5) and S (2, 7, −1).

(Solution) a) $|\mathbf{u}| = u = \sqrt{3^2 + 4^2 + (-7)^2} = \sqrt{9 + 16 + 49} = \sqrt{74}$

Note that you can simply write u for the length of **u**, but **only** if you are meticulous about writing $\underline{u}$ when you mean the vector. You can avoid such subtlety and possible confusion by using the $|\underline{u}|$ notation and never writing your vectors without their underscores.

b) $RS = \sqrt{(3-2)^2 + (-4-7)^2 + (5-(-1))^2} = \sqrt{1 + 121 + 36} = \sqrt{158}$

Remember to add the squares. Remember that $(-11)^2$ is not −121.

13·8 how to determine whether or not three points in space are collinear, and, if collinear, to determine the ratio in which one divides the join of the other two

Example

Show that the points K$(-2, -3, 5)$, L$(2, 3, 3)$ and M$(8, 12, 0)$ are collinear and find the ratio in which L divides KM.
(This might be worth 4 marks in the exam.)

(Solution) The method consists of finding two parallel line segments with a point in common. Any two of the six possible line segments will do, but since we are asked to find the ratio KL : LM, it is most sensible to find $\overrightarrow{KL}$ and $\overrightarrow{LM}$.

$$\overrightarrow{KL} = \mathbf{l} - \mathbf{k} = \begin{pmatrix} 2 \\ 3 \\ 3 \end{pmatrix} - \begin{pmatrix} -2 \\ -3 \\ 5 \end{pmatrix} = \begin{pmatrix} 4 \\ 6 \\ -2 \end{pmatrix}$$ (the first mark)

$$\overrightarrow{LM} = \mathbf{m} - \mathbf{l} = \begin{pmatrix} 8 \\ 12 \\ 0 \end{pmatrix} - \begin{pmatrix} 2 \\ 3 \\ 3 \end{pmatrix} = \begin{pmatrix} 6 \\ 9 \\ -3 \end{pmatrix} = \frac{3}{2}\begin{pmatrix} 4 \\ 6 \\ -2 \end{pmatrix} = \frac{3}{2}\overrightarrow{KL}$$ (the second mark)

$\Rightarrow$ LM || KL and L is common to both $\Rightarrow$ K, L and M are collinear
(the third mark)

also $\overrightarrow{LM} = \frac{3}{2}\overrightarrow{KL} \Rightarrow \frac{\text{KL}}{\text{LM}} = \frac{2}{3}$ [notice that we cannot divide vectors]
or KL : LM = 2 : 3 (the fourth mark)

Compare this proof of collinearity with that in item 3·7. We could have used two dimensional vectors there, but **remember** that there is NO such thing as a GRADIENT in three dimensions.

Note also that by applying item 13·3,
$\overrightarrow{AB} = k\overrightarrow{CD} \Rightarrow$ AB||CD but this does not necessarily mean that A, B, C, D are collinear.

Although it may be obvious above that L is a common point, do not omit reference to this crucial aspect of the proof.

13·9 how to calculate the co-ordinates of a point which divides the line joining two given points in a given ratio (preferably by using the section formula)

The Section Formula: $\mathbf{p} = \dfrac{m\mathbf{b} + n\mathbf{a}}{m + n}$

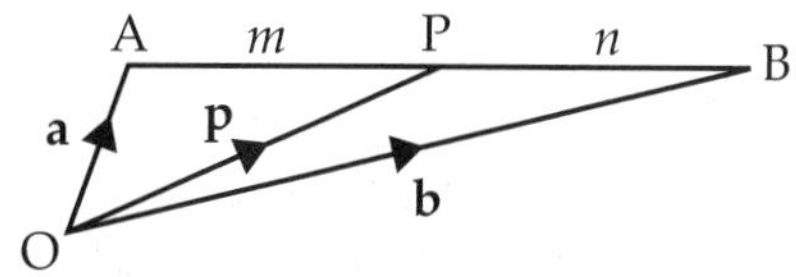

This is also often written as $\mathbf{p} = \dfrac{1}{m+n}(m\mathbf{b} + n\mathbf{a})$

Example A is the point (3, 4, 5) and B (8, −6, 15).
Find the coordinates of the point P which divides AB in the ratio 3 : 2.

(Solution) Make a quick sketch every time, even with the line straight across the page as shown, adding in the crossing arrows from the '3' and the '2' to avoid errors with which multiplier is attached to which position vector.

$$\mathbf{p} = \frac{3\begin{pmatrix}8\\-6\\15\end{pmatrix} + 2\begin{pmatrix}3\\4\\5\end{pmatrix}}{3+2} = \frac{1}{5}\begin{pmatrix}30\\-10\\55\end{pmatrix} = \begin{pmatrix}6\\-2\\11\end{pmatrix}$$

(3) P (2)
A B
(3, 4, 5) (8, −6, 15)

i.e. P is (6, −2, 11)

(alternative solution) some pupils (and perhaps their teachers) prefer to do this kind of example from first principles every time, as follows:

$$\overrightarrow{AB} = \mathbf{b} - \mathbf{a} = \begin{pmatrix}8\\-6\\15\end{pmatrix} - \begin{pmatrix}3\\4\\5\end{pmatrix} = \begin{pmatrix}5\\-10\\10\end{pmatrix}$$

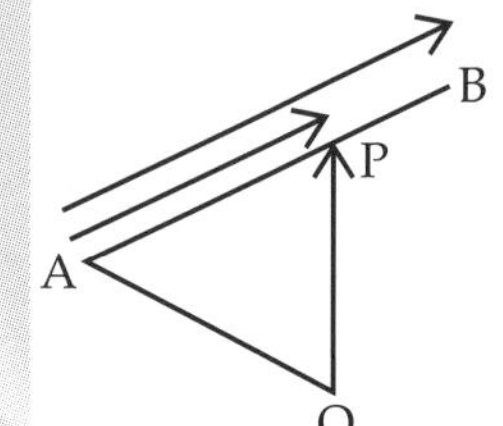

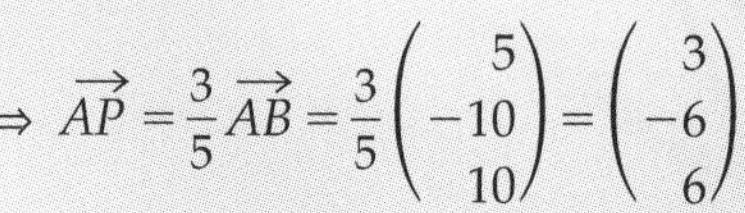

$$\Rightarrow \overrightarrow{AP} = \frac{3}{5}\overrightarrow{AB} = \frac{3}{5}\begin{pmatrix}5\\-10\\10\end{pmatrix} = \begin{pmatrix}3\\-6\\6\end{pmatrix}$$

$$\Rightarrow \mathbf{p} = \overrightarrow{OP} = \overrightarrow{OA} + \overrightarrow{AP} = \begin{pmatrix}3\\4\\5\end{pmatrix} + \begin{pmatrix}3\\-6\\6\end{pmatrix} = \begin{pmatrix}6\\-2\\11\end{pmatrix}, \text{ hence P is } (6, -2, 11)$$

(In principle, I am all in favour of doing mathematics with the minimum of formulae (unless the user is capable of giving a proof) and approve of this second approach, but in practice I see too many scripts where the third step is omitted and the pupil seems to think that P should be (3, −6, 6).

The section formula avoids this pitfall and is also more useful if you encounter examples using letters instead of numbers. Just make sure you apply the correct ratio in your diagram.)

13·10 that the scalar product is defined by $\mathbf{a.b} = |\mathbf{a}||\mathbf{b}|\cos\theta$ where θ is the angle between the positive directions of a and b

$\mathbf{a.b} = a \times b \times \cos\theta$ or $\mathbf{a.b} = |\mathbf{a}| \times |\mathbf{b}| \times \cos\theta$
a is the vector
a is the length of **a**

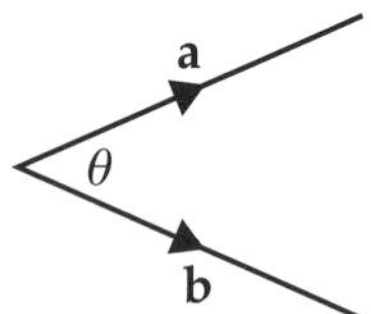

Remember that **a.b** is a number, not a vector. This is why it is called the scalar product.

Do not use × instead of a dot. The × is an arithmetic multiplier but means something else in Advanced Higher Maths vector work.

Be careful with θ. It must be the angle between the positive directions of the vectors, i.e. both vectors issuing from the same point or both vectors entering the same point.

Example

Find the value of a) **u.v**
b) **p.q**.

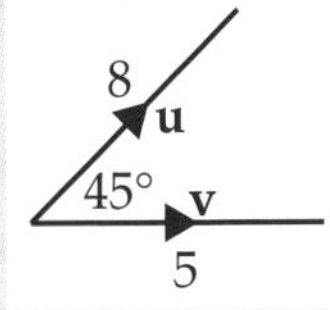

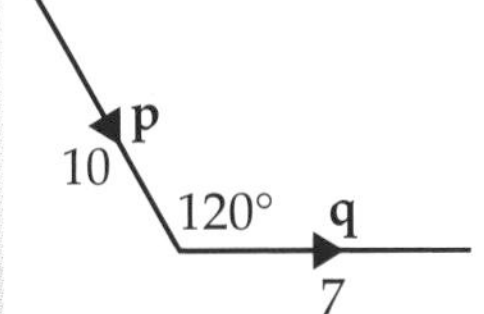

(Solution) a) $\mathbf{u.v} = |\mathbf{u}| \times |\mathbf{v}| \times \cos\theta = 8 \times 5 \times \cos 45° = 40 \times \frac{1}{\sqrt{2}} = 20 \times 2 \times \frac{1}{\sqrt{2}} = 20\sqrt{2}$

b) 120° is not the angle between the positive directions of **p** and **q**.
It is worthwhile making another sketch:
Draw **p** and **q** 'coming out of the same point'.
This shows that the value of θ is 60°.
[It would be equally valid for **p** and **q** to be 'going in to the same point'.]

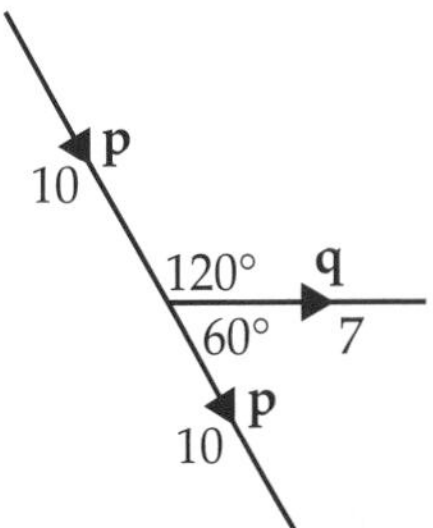

hence $\mathbf{p.q} = |\mathbf{p}| \times |\mathbf{q}| \times \cos\theta = 10 \times 7 \times \cos 60°$

$= 70 \times \left(\frac{1}{2}\right) = 35$

13·11 that a.b = $a_1b_1 + a_2b_2 + a_3b_3$ also defines the scalar product

The approach to the scalar product in item 13·10 is essentially geometrical. This result is more algebraic. It is used when the vectors are known in component form.

Example

Find the value of **r.s** where $\mathbf{r} = 3\mathbf{i} - 4\mathbf{j} + 7\mathbf{k}$ and $\mathbf{s} = 2\mathbf{i} - 3\mathbf{j} - 5\mathbf{k}$.

(Solution) $\mathbf{r} = 3\mathbf{i} - 4\mathbf{j} + 7\mathbf{k} = \begin{pmatrix} 3 \\ -4 \\ 7 \end{pmatrix}$ $\quad \mathbf{s} = 2\mathbf{i} - 3\mathbf{j} - 5\mathbf{k} = \begin{pmatrix} 2 \\ -3 \\ -5 \end{pmatrix}$

$\mathbf{r.s} = 3 \times 2 + [(-4) \times (-3)] + [7 \times (-5)] = 6 + 12 - 35 = -17$

It was not absolutely necessary to introduce column vectors, but most people are happier using them than the basis unit vectors **i**, **j**, **k**.

13·12 how to calculate the angle between two vectors (given in component form)

The method is to use the two equivalent forms of the scalar product (given in items 13·10 and 13·11) to find the cosine of the angle and hence the angle. A question on this in the calculator paper is almost certain in your final exam. Casual errors at any point seldom prevent you from carrying out further relevant working, so try to keep going to the end of your calculation. The most common place for errors is in questions where you have to interpret a diagram to establish some co-ordinates first of all. Take care with this easy bit at the start, but remember that even if you get these wrong, you can still score full marks on the calculation of the angle. Remember too that if the cosine of the angle turns out to be negative, then the angle is obtuse.

Example

Calculate the angle between $\mathbf{r}$ and $\mathbf{s}$ where $\mathbf{r} = 3\mathbf{i} - 4\mathbf{j} + 7\mathbf{k}$ and $\mathbf{s} = 2\mathbf{i} - 3\mathbf{j} - 5\mathbf{k}$.

(Solution) $\mathbf{r.s} = 3 \times 2 + [(-4) \times (-3)] + [7 \times (-5)] = 6 + 12 - 35 = -17$ (as in item 13·11)

also $|\mathbf{r}| = \sqrt{3^2 + (-4)^2 + 7^2} = \sqrt{9 + 16 + 49} = \sqrt{74}$

and $|\mathbf{s}| = \sqrt{2^2 + (-3)^2 + (-5)^2} = \sqrt{4 + 9 + 25} = \sqrt{38}$

Let θ be the required angle, then

$$\mathbf{r.s} = |\mathbf{r}| \times |\mathbf{s}| \times \cos\theta \Rightarrow \cos\theta = \frac{\mathbf{r.s}}{|\mathbf{r}| \times |\mathbf{s}|}$$

$$\text{hence } \cos\theta = \frac{-17}{\sqrt{74} \times \sqrt{38}} \Rightarrow \theta = 180 - 71{\cdot}3 = 108{\cdot}7^\circ$$

13·13 that the scalar product is distributive over addition, ie. a.(b + c) = a.b + a.c

Example

Find the value of $\mathbf{v}.(\mathbf{v} + \mathbf{w})$ where $|\mathbf{v}| = 3$, $|\mathbf{w}| = 4$ and the angle between $\mathbf{v}$ and $\mathbf{w}$ is 60°.

(Solution) Calculating the angle between $\mathbf{v}$ and $(\mathbf{v} + \mathbf{w})$ is not an easy option. It requires the cosine rule, the sine rule, and the use of $\cos^2 x + \sin^2 x = 1$ and I have never seen anyone complete this working under exam conditions. It takes far too long anyway, so the recommended way (and easiest way) is to use the distributive law, as follows:

$$\begin{aligned}\mathbf{v}.(\mathbf{v} + \mathbf{w}) &= \mathbf{v.v} + \mathbf{v.w} \\ &= (3 \times 3 \times \cos 0^\circ) + (3 \times 4 \times \cos 60^\circ) \\ &= 9 \times 1 + 12 \times \frac{1}{2} \\ &= 15\end{aligned}$$

13·14 that for a, b $\neq 0$, $\mathbf{a}.\mathbf{b} = 0 \Leftrightarrow$ a is perpendicular to b.

$\mathbf{a}.\mathbf{b} = 0 \Leftrightarrow |\mathbf{a}| \times |\mathbf{b}| \times \cos\theta = 0 \Leftrightarrow \cos\theta = 0$ because neither $|\mathbf{a}|$ nor $|\mathbf{b}|$ is 0
so $\theta = 90°$ i.e. **a** and **b** are perpendicular

Example

Show that the vectors **s** and **t** are perpendicular to each other where $\mathbf{s} = 2\mathbf{i} + 3\mathbf{j} - \mathbf{k}$ and $\mathbf{t} = 5\mathbf{i} - \mathbf{j} + 7\mathbf{k}$.

(Solution) $\mathbf{s} = 2\mathbf{i} + 3\mathbf{j} - \mathbf{k} = \begin{pmatrix} 2 \\ 3 \\ -1 \end{pmatrix}$ $\mathbf{t} = 5\mathbf{i} - \mathbf{j} + 7\mathbf{k} = \begin{pmatrix} 5 \\ -1 \\ 7 \end{pmatrix}$

[$\mathbf{s}.\mathbf{t} = 0$ for **s** and **t** to be perpendicular to each other, so calculate **s.t**]

$\mathbf{s}.\mathbf{t} = (2 \times 5) + (3 \times (-1)) + ((-1) \times 7)$ **(not essential working – only here as explanation)**

$= 10 - 3 - 7 = 0$

hence **s** is perpendicular to **t**.

[It is not absolutely essential to write **s** and **t** as column vectors, but if you write $\mathbf{s}.\mathbf{t} = (2\mathbf{i} + 3\mathbf{j} - \mathbf{k}).(5\mathbf{i} - \mathbf{j} + 7\mathbf{k})$, strictly speaking when you expand the brackets there should be nine terms, six of which turn out to be zero.]

13·15 how to use vectors to justify geometrical results and apply vectors in problem situations

(See Chapter 18 on typical longer exam questions:
Questions 18·8 and 18·12. Look back also at item 13·13.)

Chapter 14

FURTHER CALCULUS

What You Should Know

14·1 that $\frac{d}{dx}(\sin x) = \cos x$ and $\frac{d}{dx}(\cos x) = -\sin x$

14·2 how to differentiate a function of a function of x using the chain rule $\frac{dy}{dx} = \frac{dy}{du} \times \frac{du}{dx}$

14·3 that $\int \sin x \, dx = -\cos x + c$ and $\int \cos x \, dx = \sin x + c$

14·4 that $\int \sin(ax + b)\, dx = -\frac{1}{a}\cos(ax + b) + c$

$$\int \cos(ax + b)\, dx = \frac{1}{a}\sin(ax + b) + c$$

$$\int (ax + b)^n \, dx = \frac{(ax + b)^{n+1}}{a(n + 1)} + c \qquad n \neq -1$$

Notes on items 14·1 to 14·4: Further Calculus

14·1 that $\frac{d}{dx}(\sin x) = \cos x$ and $\frac{d}{dx}(\cos x) = -\sin x$

Example

Differentiate $2x^2 + 3\sin x + 4\cos x$ with respect to x.

(Solution) $\frac{d}{dx}(2x^2 + 3\sin x + 4\cos x) = 4x + 3\cos x - 4\sin x$

14·2 how to differentiate a function of a function of x using the chain rule $\frac{dy}{dx} = \frac{dy}{du} \times \frac{du}{dx}$

You need to be able to use the chain rule to differentiate a function of a function. We can think of there being three types:

a bracket to a power

a bracket involving a trig function to a power

a trig function of a function of x

(All of this item is above grade C, but C grade candidates should be able to score some of the available marks.)

Example

Differentiate a) $(2x^2 - 1)^{10}$ b) $\sin^3 x$ c) $\cos(2x^3 - 3x + 5)$.

(Solution) In general, we think of peeling an onion, one skin at a time.

a) (Think of '(something)^{10}'. The derivative of u^{10} is $10u^9$.
So think '$10 \times$ (something)$^9 \times$ the derivative of the (something)'.)

$$\frac{d}{dx}(2x^2-1)^{10} = 10 \times (2x^2-1)^9 \times \frac{d}{dx}(2x^2-1)$$
$$= 10 \times (2x^2-1)^9 \times 4x$$
$$= 40x(2x^2-1)^9$$

b) (Think of $\sin^3 x$ as $(\sin x)^3$. The derivative of u^3 is $3u^2$.
So think '$3 \times$ (the bracket)$^2 \times$ the derivative of (the bracket)'.)

$$\frac{d}{dx}(\sin^3 x) = \frac{d}{dx}(\sin x)^3 = 3(\sin x)^2 \times \frac{d}{dx}(\sin x)$$
$$= 3(\sin x)^2 \times \cos x$$
$$= 3\sin^2 x \cos x$$

c) (Think of 'cos (something)'. The derivative of $\cos(u)$ is $-\sin(u)$.
So think '$-\sin$ (something) $\times$ the derivative of the (something)'.)

$$\frac{d}{dx}\cos(2x^3-3x+5) = -\sin(2x^3-3x+5) \times \frac{d}{dx}(2x^3-3x+5)$$
$$= -\sin(2x^3-3x+5) \times (6x^2-3)$$
$$= 3(1-2x^2)\sin(2x^3-3x+5)$$

14·3 that $\int \sin x \, dx = -\cos x + c$ and $\int \cos x \, dx = \sin x + c$

Example

Find $\int (3x^2 - 4\cos x + 5\sin x)\, dx$.

(Solution) $\int (3x^2 - 4\cos x + 5\sin x)\, dx = x^3 - 4\sin x - 5\cos x + c$ [Differentiate to check]

! **An easy way to remember these differentiations and integrations of trig functions is to learn thoroughly that <u>the derivative of $\sin x$ is $\cos x$</u>. From this you can deduce that <u>the integral of $\cos x$ is $\sin x$</u>. So it must be the other two results that involve a negative sign, and the function changes (from sin to cos and vice versa) every time. <u>Learn these rules</u>.**

14·4 **that** $\int \sin(ax+b)\,dx = -\frac{1}{a}\cos(ax+b) + c,$

$$\int \cos(ax+b)\,dx = \frac{1}{a}\sin(ax+b) + c,$$

$$\int (ax+b)^n\,dx = \frac{(ax+b)^{n+1}}{a(n+1)} + c \qquad n \neq -1$$

The 'a' on the bottom line is the derivative of $(ax+b)$. These results can be proved by differentiating the right hand sides, using the chain rule. These are three special integrals, which apply only to the integration of functions of linear functions, i.e. these rules do not apply, for example, to the integrals of $(3x^2+5)^4$ or $\sin(2x^2)$. Again learn these rules.

Example

Find a) $\int \cos(3x+4)\,dx$ b) $\int_0^{\frac{\pi}{2}} \sin(2x)\,dx$ c) $\int_2^6 \frac{1}{\sqrt{2x-3}}\,dx$

(Solution) a) $\int \cos(3x+4)\,dx = \frac{1}{3}\sin(3x+4) + c$ (differentiate the answer as a check)

b) $\int_0^{\frac{\pi}{2}} \sin(2x)\,dx = \left[\frac{-\cos 2x}{2}\right]_0^{\frac{\pi}{2}} = -\frac{1}{2}[\cos\pi - \cos 0] = -\frac{1}{2}[-1-1] = 1$

(Notice how I took out the common factor of $-\frac{1}{2}$ before substituting; this generally leads to less error. Note also that the value of $\cos(2x)$ when x is zero is NOT zero.)

c) $\int_2^6 \frac{1}{\sqrt{2x-3}}\,dx = \int_2^6 \frac{1}{(2x-3)^{\frac{1}{2}}}\,dx = \int_2^6 (2x-3)^{-\frac{1}{2}}\,dx$ (use indices carefully)

$$= \left[\frac{(2x-3)^{\frac{1}{2}}}{\frac{1}{2}\times 2}\right]_2^6 = (2\times 6-3)^{\frac{1}{2}} - (2\times 2-3)^{\frac{1}{2}}$$

$$= 9^{\frac{1}{2}} - 1^{\frac{1}{2}} = 3 - 1 = 2$$

(Questions **18·13** and **18·14** (in Chapter 18) are also on this topic. They are not particularly long but would appear near the end of an exam paper.

Look back also at item **14·2**.)

Chapter 15

LOGARITHMS

What You Should Know

15·1 that $N = b^x \Leftrightarrow \log_b N = x\ (b > 1, x > 0)$ and hence that for all b, $\log_b b = 1$ and $\log_b 1 = 0$

15·2 the general features of the graphs of $f(x) = \log_b x$ and $f(x) = a^x$; that the domain of $f(x) = \log_a x$ is $\{x \in \mathbb{R} : x > 0\}$, and that $\log_a x = \log_a y \Leftrightarrow x = y$

15·3 the laws of logarithms:
I $\log_a uv = \log_a u + \log_a v$
II $\log_a\left(\frac{u}{v}\right) = \log_a u - \log_a v$
III $\log_a u^v = v \times \log_a u$
and how to use them to manipulate logarithmic expressions

15·4 how to sketch associated graphs, e.g. $\log(x-1)$, $\log\frac{1}{x}$, $\log(2x)$

15·5 how to determine any one of a, b, c given the other two in the equation $\log_b a = c$, or $a = b^c$

15·6 how to solve for a and b equations of the following forms, (given two pairs of corresponding values of x and y): $\log y = a \log x + b$, $y = ax^b$, $y = ab^x$

15·7 how to verify that data are consistent with relationships of the form $y = ax^n$ or $y = ab^x$

Notes on items 15·1 to 15·7: Logarithms

15·1 that $N = b^x \Leftrightarrow \log_b N = x\ (b > 1, x > 0)$ and hence that for all b, $\log_b b = 1$ and $\log_b 1 = 0$

(This definition of a logarithm was first met in item 4·7, where the log function was defined as the inverse of the exponential function and the sketch of their graphs showed them as images of each other under reflection in the line with equation $y = x$.)

This definition applied to $a^1 = a$ gives $\log_a a = 1$, and to $a^0 = 1$ gives $\log_a 1 = 0$. Hence the points $(1, 0)$ and $(a, 1)$ lie on the graph of $y = \log_a x$.

(You must be able to sketch this graph instantly.)

(Other standard graphs which should also be at your finger tips include: $y = a^x$, $y = x^2$, $y = x^3$ as well as many linear functions.)

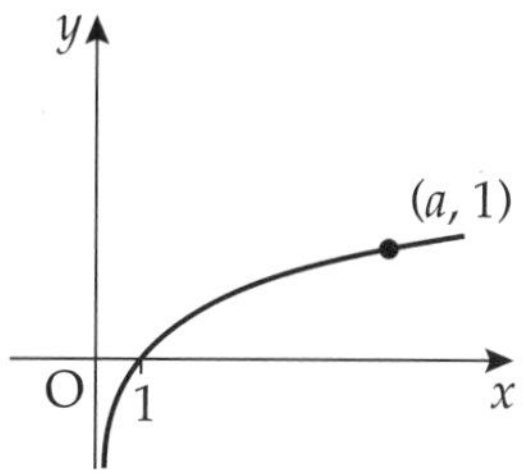

15·2 the general features of the graphs of $f(x) = \log_b x$ and $f(x) = a^x$; that the domain of $f(x) = \log_a x$ is $\{x \in \mathbb{R} : x > 0\}$, and that $\log_a x = \log_a y \Leftrightarrow x = y$

These graphs appeared in item 4·7. In item 15·1, the graph of $y = \log_a x$ shows that y is only defined for $x > 0$.

This is the same as saying that the domain of $f(x) = \log_a x$ is $\{x \in \mathbb{R} : x > 0\}$.

Note that the graph is never at the same height twice and that the range is the whole of $\mathbb{R}$.

These properties make it a one-to-one correspondence, which was introduced in item 4·2.

Therefore, if two numbers are equal, then their logarithms are equal, and if the logarithms of two numbers are equal,then the numbers themselves are equal. This is concisely expressed by the mathematical statement that $\log_a x = \log_a y \Leftrightarrow x = y$.

15·3 the laws of logarithms

Remember

I $\log_a uv = \log_a u + \log_a v$

II $\log_a\left(\frac{u}{v}\right) = \log_a u - \log_a v$

III $\log_a u^v = v \times \log_a u$

Here we have another example of formulae which must be known in both directions. Sometimes, for example, we might want to change $\log 6$ into $(\log 2 + \log 3)$, but in different circumstances we might want to change $(\log 2 + \log 3)$ into $\log 6$. Omitting the base indicates that the statement is valid for any chosen base.

Example

Express $2 + \log_{10}42 - \log_{10}28$ as the logarithm of a single number.

(Solution) Since, $\log_{10}10 = 1, 2 = 2 \times 1 = 2 \times \log_{10}10 = \log_{10}100$ (using the 3rd law)

$\therefore 2 + \log_{10}42 - \log_{10}28 = \log_{10}100 + \log_{10}\left(\frac{42}{28}\right)$ (using the 2nd law)

$= \log_{10}\left(100 \times \frac{3}{2}\right) = \log_{10}150$ (using the 1st law)

Example

Solve $\log_3(2x) - \log_3(x+1) = -1$.

(Solution) $\log_3(2x) - \log_3(x+1) = -1$

$\Rightarrow \log_3\left(\frac{2x}{x+1}\right) = -1$ (using the 2nd law)

$\Rightarrow \left(\frac{2x}{x+1}\right) = 3^{-1} = \frac{1}{3}$ (change from log to index form)

$\Rightarrow 6x = x+1 \Rightarrow 5x = 1 \Rightarrow x = \frac{1}{5}$ (cross multiply & solve)

Example

Express $\log_{14}\left(\frac{63}{8}\right)$ a) in terms of $\log_{14}2$, $\log_{14}3$ and $\log_{14}7$, and hence
b) in terms of $\log_{14}2$ and $\log_{14}3$.

(Solution) a) $\log_{14}\left(\frac{63}{8}\right) = \log_{14}63 - \log_{14}8$ (using the 2nd law)

$= \log_{14}(7 \times 9) - \log_{14}8$

$= \log_{14}7 + \log_{14}9 - \log_{14}8$ (using the 1st law)

$= \log_{14}7 + \log_{14}3^2 - \log_{14}2^3$

$= \log_{14}7 + 2\log_{14}3 - 3\log_{14}2$ (using the 3rd law)

b) log 9 was easily expressed in terms of log 3, as was log 8 in terms of log 2 but we need a little trick to deal with log 7; this is the slightly harder but notice that $7 = 14 \div 2$, so to simplify $\log_{14}7$ we proceed as follows:

$\log_{14}7 = \log_{14}\left(\frac{14}{2}\right) = \log_{14}14 - \log_{14}2$ (using the 2nd law)

$= 1 - \log_{14}2$ (using $\log_a a = 1$)

hence $\log_{14}\left(\frac{63}{8}\right) = \log_{14}7 + 2\log_{14}3 - 3\log_{14}2$ (from part (a))

$= 1 - \log_{14}2 + 2\log_{14}3 - 3\log_{14}2$

$= 1 + 2\log_{14}3 - 4\log_{14}2$ (gather like terms)

15·4 how to sketch associated graphs

Example

Sketch the graphs of a) $y = \log_5(x-1)$

b) $y = \log_a\frac{1}{x}$

c) $y = \log_a(2x)$

***Example** continued* ➢

***Example** continued*

(Solution) a) The $(x - 1)$ tells us that this is a translation of 1 unit to the right of the graph of $y = \log_5(x)$, so it crosses the x-axis at (2, 0) and passes through (6, 1).

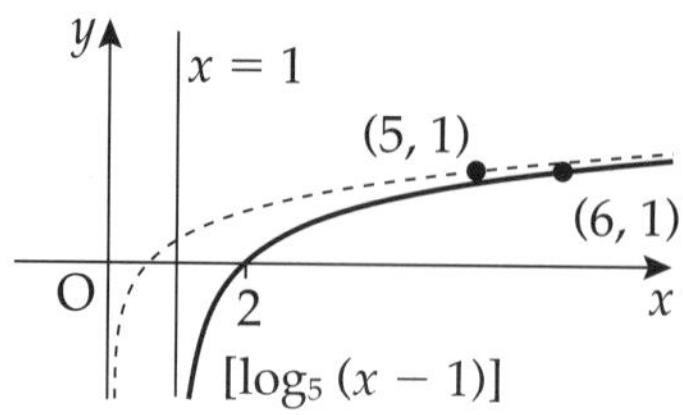

The asymptote $x = 0$ moves too, to become $x = 1$.

(This example is straightforward and does not require the application of the laws of logs. It could have been included in item **4·8**.)

b) If $f(x) = \log_a x$, then $y = \log_a \frac{1}{x} = \log_a(x^{-1}) = -\log_a x = -f(x)$ so reflect the graph of $y = \log_a x$ in the x-axis [see item **4·8**]; it will pass through (1, 0) and $(a, -1)$.

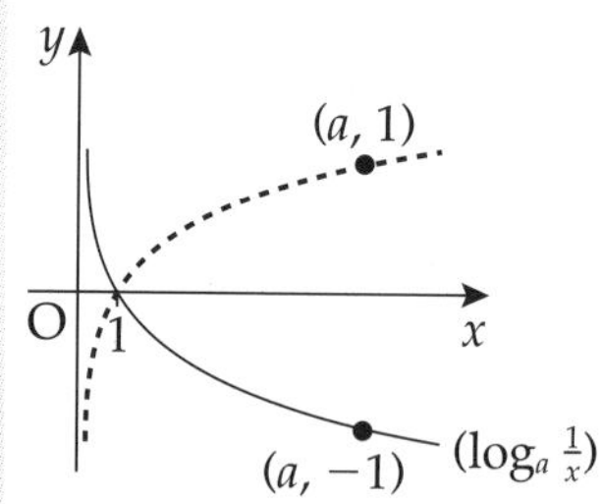

c) There are two correct ways to think of this graph:

i) item **4·8** indicates a side to side shrinking by half

ii) the first law of logs implies that $\log 2x = \log 2 + \log x$ which suggests a translation up of log 2 using either interpretation, $x = \frac{1}{2}$ gives $y = 0$ and $x = \frac{a}{2}$ gives $y = 1$

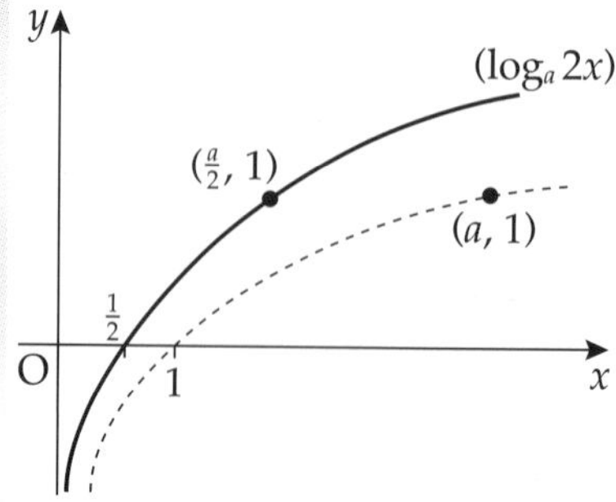

15·5 how to determine any one of a, b, c given the other two in the equation $\log_b a = c$ or $a = b^c$.

Example

Solve a) $\log_{2\cdot 1} x = 3\cdot 6$ b) $3\cdot 6 = 2\cdot 1^x$ c) $3\cdot 6 = x^{2\cdot 1}$.

(Solution) a) Change to the index form, and use a calculator. $x = 2\cdot 1^{3\cdot 6} = 14\cdot 45$

b) take the log of both sides: $\log 3\cdot 6 = \log 2\cdot 1^x$
use the third law of logs: $\log 3\cdot 6 = x \log 2\cdot 1$
make x the subject: $x = \dfrac{\log 3\cdot 6}{\log 2\cdot 1}$
use either $\log_{10}$ or $\log_e$ (i.e. $\ln$) consistently and so $x = 1\cdot 726$

c) this time decide whether to use $\log_{10}$ or $\log_e$ from the first line; either would do, but the working must be consistent
$\ln 3\cdot 6 = \ln(x^{2\cdot 1}) = 2\cdot 1 \ln x$

[The $\ln$ notation is never used in exam papers even though it appears on nearly everybody's calculator.]

Thus $\ln x = \dfrac{\ln 3\cdot 6}{2\cdot 1} = 0\cdot 610$ Hence $x = e^{0\cdot 610} = 1\cdot 84$

15·6 how to solve for a and b equations of the following forms (given two pairs of corresponding values of x and y): $\log y = a \log x + b$, $y = ax^b$, $y = ab^x$

Of all the syllabus items listed by S.Q.A., I consider this the least likely to appear in the actual exam. Since nothing is certain, I will include one example for you, but make sure you know everything else in this book first before you spend any time here

Example

The variables x and y satisfy a relationship of the form $y = ax^b$.
When x is 2, y is 80, and when x is 3, y is 405.
Find the values of a and b.

(Solution) $(2, 80) \Rightarrow 80 = a(2)^b$
$(3, 405) \Rightarrow 405 = a(3)^b$
eliminate a by dividing these equations $\Rightarrow \dfrac{405}{80} = \dfrac{a \times 3^b}{a \times 2^b}$
$\Rightarrow 5\cdot 0625 = 1\cdot 5^b$
the unknown is now in the index, so take the $\log_e$ (or $\log_{10}$) of both sides
i.e. $\log(5\cdot 0625) = \log(1\cdot 5)^b = b\log(1\cdot 5)$
$\Rightarrow b = \dfrac{\log(5\cdot 0625)}{\log(1\cdot 5)} = 4 \Rightarrow 80 = a \times 2^4 = 16a \Rightarrow a = 5$

[The perceptive reader will have noticed that $\dfrac{405}{80} = \dfrac{81}{16} = \left(\dfrac{3}{2}\right)^4$ hence $b = 4$]

15·7 how to verify that data are consistent with relationships of the form $y = ax^n$ or $y = ab^x$

Example

a) Show that if the variables p and q are related by an equation of the form $p = aq^b$, then there is a linear relationship between P and Q where $P = \log p$ and $Q = \log q$.

b) An experiment is conducted to test the hypothesis that the variables p and q are related by an equation of the form $p = aq^b$. It is found that there is a linear relationship between P and Q where $P = \log_{10} p$ and $Q = \log_{10} q$. The graph is shown. Find the equation relating p and q.

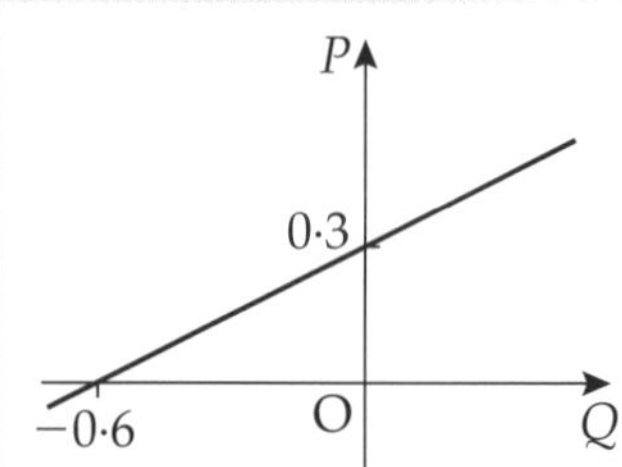

(Solution) a) $p = aq^b \Leftrightarrow \log(p) = \log(aq^b)$ (take logs of both sides)

$\Leftrightarrow \log(p) = \log(a) + \log(q^b)$ (first law of logs)

$\Leftrightarrow \log(p) = \log(a) + b\log(q)$ (third law of logs)

i.e. $P = c + bQ$ which is a linear relationship (where $c = \log(a)$)
(Notice that the process is completely reversible.)

b) we need to find the equation of the line, in terms of P and Q

$m = \frac{0{\cdot}3}{0{\cdot}6} = \frac{1}{2}$; using $y = mx + c \Rightarrow P = \frac{1}{2}Q + 0{\cdot}3$

substitute the log expressions $\Rightarrow \log_{10} p = \frac{1}{2}\log_{10} q + 0{\cdot}3$

Method 1 (keep the 0·3 as it is and manipulate the other terms)

$\log_{10} p = \frac{1}{2}\log_{10} q + 0{\cdot}3 \Rightarrow \log_{10} p - \log_{10} q^{\frac{1}{2}} = 0{\cdot}3$ (third law of logs)

$\Rightarrow \log_{10}\left(\frac{p}{\sqrt{q}}\right) = 0{\cdot}3$ (second law of logs)

$\Rightarrow \frac{p}{\sqrt{q}} = 10^{0{\cdot}3} = 2$ (log to index form)

$\Rightarrow p = 2\sqrt{q}$

Method 2 (we need to write 0·3 as a $\log_{10}$(something))

suppose $0{\cdot}3 = \log_{10} S$, then $10^{0{\cdot}3} = S = 2 \Rightarrow 0{\cdot}3 = \log_{10} 2$

so $\log_{10} p = \frac{1}{2}\log_{10} q + 0{\cdot}3 \Rightarrow \log_{10} p = \frac{1}{2}\log_{10} q + \log_{10} 2$

$\Rightarrow \log_{10} p = \log_{10} q^{\frac{1}{2}} + \log_{10} 2$ (third law of logs)

$\Rightarrow \log_{10} p = \log_{10} 2q^{\frac{1}{2}}$ (first law of logs)

hence $p = 2\sqrt{q}$

(Question 18·9 in Chapter 18 is also on this topic. Look back also at item 15·7.)

Chapter 16

WAVE FUNCTION

THE ~~AUXILIARY ANGLE~~

What You Should Know

16·1 how to solve a pair of equations of the form $\begin{cases} k\cos\alpha = a \\ k\sin\alpha = b \end{cases}$

16·2 how to express $a\cos x + b\sin x$ (a, b given constants) in the form $k\cos(x \pm \theta)$ and $k\sin(x \pm \theta)$ with $k > 0$ and θ in a specified range (usually $0 < \theta < 2\pi$, or in degrees throughout), and that θ is called the auxiliary angle

16·3 how to use the auxiliary angle to find zeros of the function $a\cos x + b\sin x$, its maximum and minimum values, and the corresponding values of x

16·4 how to use the auxiliary angle to sketch the graph of $y = a\cos x + b\sin x$

16·5 how to solve equations of the form $a\cos x + b\sin x = c$

16·6 how to deal with $a\cos px + b\sin px$.

Notes on items 16·1 to 16·6: The Auxiliary Angle

16·1 how to solve a pair of equations of the form $\begin{cases} k\cos\alpha = a \\ k\sin\alpha = b \end{cases}$

Example Solve $\begin{cases} k\cos\alpha = 5 \\ k\sin\alpha = -12 \end{cases}$ for $k > 0$ and $0 < \alpha < 360$.

(Solution) (Note that you have to make use of both the trig identities mentioned in item 7·1.)

squaring both equations: $k^2\cos^2\alpha = 25$

$k^2\sin^2\alpha = 144$

adding: $k^2\cos^2\alpha + k^2\sin^2\alpha = 169$

$\Rightarrow k^2(\cos^2\alpha + \sin^2\alpha) = 169$

$\Rightarrow k = 13$ (since $\cos^2\alpha + \sin^2\alpha = 1$)

dividing the equations: $\dfrac{k\sin\alpha^\circ}{k\cos\alpha^\circ} = -\dfrac{12}{5} \Rightarrow \tan\alpha = -2{\cdot}4$

$k > 0 \Rightarrow \cos\alpha^\circ > 0 \Rightarrow \alpha$ lies in the first or fourth quadrants $k > 0 \Rightarrow \sin\alpha^\circ < 0 \Rightarrow \alpha$ lies in the third or fourth quadrants hence α lies in the fourth quadrant $\Rightarrow \alpha = 292{\cdot}6^\circ$

sin	All ✓
✓ tan	✓ ✓ cos

16·2 **how to express $a\cos x + b\sin x$ (a, b, given constants) in the form $k\cos(x \pm \theta)$ and $k\sin(x \pm \theta)$ with $k > 0$ and θ in a specified range (usually $0 < \theta < 2\pi$, or in degrees throughout), and that θ is called the auxiliary angle.**

Carelessness and lack of communication are the usual causes of loss of marks in the basic type of question. There are two common methods. You should stick to the method your teacher taught you. I shall begin with the more popular.

Example

Express $2\sqrt{3}\cos x° - 5\sqrt{2}\sin x°$ as a single trig function.
(You may alternatively be asked to write it in the form $R\cos(x+\alpha)°$ where $R > 0$ and $0 < \alpha < 360$.)

(Solution) Method 1

Let $2\sqrt{3}\cos x° - 5\sqrt{2}\sin x° = k\cos(x+\alpha)°$
i.e. $2\sqrt{3}\cos x° - 5\sqrt{2}\sin x° = k(\cos x°\cos\alpha° - \sin x°\sin\alpha°)$
$\Rightarrow 2\sqrt{3}\cos x° - 5\sqrt{2}\sin x° = k(\cos\alpha°)\cos x° - (k\sin\alpha°)\sin x°$

(this earns the first mark – expanding the compound angle correctly)

equating coefficients of $\cos x°$: $2\sqrt{3} = k\cos\alpha°$
equating coefficients of $\sin x°$: $-5\sqrt{2} = -k\sin\alpha°$ i.e. $5\sqrt{2} = k\sin\alpha°$

(DO NOT omit '*k*' from these equations or they become meaningless.)
(this earns the second mark – obtaining, with justification, this pair of equations)
to find k, square both equations and add, then use $\sin^2 x° + \cos^2 x° = 1$

$k^2\cos^2\alpha° = (2\sqrt{3})^2 = 12$
$k^2\sin^2\alpha° = (5\sqrt{2})^2 = 50$
adding gives $k^2\cos^2\alpha° + k^2\sin^2\alpha° = 12 + 50 = 62$
hence $k^2(\sin^2\alpha° + \cos^2\alpha°) = 62 \Rightarrow k^2 = 62$ so $k = \sqrt{62}$

(this earns the third mark – finding k)

we eliminate k and find α by dividing the equations

$$\frac{k\sin\alpha°}{k\cos\alpha°} = \frac{5\sqrt{2}}{2\sqrt{3}} \Rightarrow \tan\alpha° = \frac{5\sqrt{2}}{2\sqrt{3}} = 2{\cdot}041$$

apply the knowledge that $\sin\alpha°$ and $\cos\alpha°$ are both positive to an 'All-sin-tan-cos' diagram, to justify that α is acute; always justify why you choose a particular quadrant.

✓ sin	✓✓ All
tan	cos ✓

thus $\alpha = 63{\cdot}9$, and $k\cos(x+\alpha)° = \sqrt{62}\cos(\alpha + 63{\cdot}9)°$

this earns the fourth mark – it must be consistent with your previous equations
(Just show the working printed within the rectangles.)

***Example** continued* ➢

Example continued

Method 2 [With this method the auxiliary angle is always acute.]
draw a right-angled triangle with shorter sides $2\sqrt{3}$ and $5\sqrt{2}$;
the hypotenuse is therefore $\sqrt{62}$ (first mark)

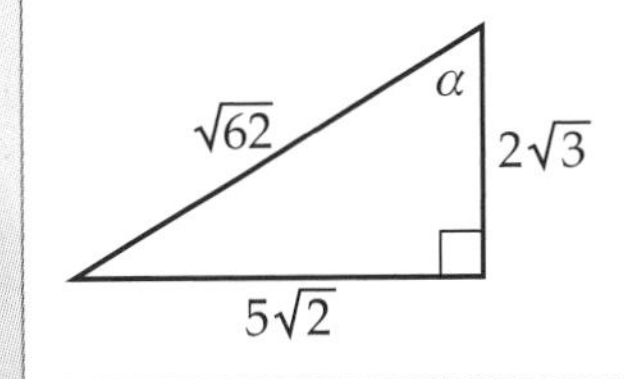

take out $\sqrt{62}$ as a common factor

$$2\sqrt{3}\cos x^\circ - 5\sqrt{2}\sin x^\circ = \sqrt{62}\left[\cos x^\circ \times \frac{2\sqrt{3}}{\sqrt{62}} - \sin x^\circ \times \frac{5\sqrt{2}}{\sqrt{62}}\right]$$ (second mark)

(to get $\cos(\ldots + \ldots)$, we need [$\cos\ldots\cos\ldots - \sin\ldots\sin\ldots$], so go back to your triangle and carefully and consistently label the appropriate acute angle α)

$$2\sqrt{3}\cos x^\circ - 5\sqrt{2}\sin x^\circ = \sqrt{62}\,[\cos x^\circ \times \cos\alpha^\circ - \sin x^\circ \times \sin\alpha^\circ]$$

where $\alpha = \tan^{-1}\left(\frac{5\sqrt{2}}{2\sqrt{3}}\right) = 63{\cdot}9$ (third mark)

hence $2\sqrt{3}\cos x^\circ - 5\sqrt{2}\sin x^\circ = \sqrt{62}\,[\cos(x^\circ + \alpha^\circ)] = \sqrt{62}\cos(x + 63{\cdot}9)^\circ$

(fourth mark)

(As with method 1, just show the working printed within the rectangles.)

16·3 how to use the auxiliary angle to find zeros of the function $a\cos x + b\sin x$, its maximum and minimum values, and the corresponding values of x

Example

Find the zeros of $f(x) = 2\sqrt{3}\cos x^\circ - 5\sqrt{2}\sin x^\circ$, the maximum and minimum values of f, and the corresponding values of x for $0 \leqslant x \leqslant 360$.

(Solution) As this is the same function as in item **16·2**, we already know that
$f(x) = \sqrt{62}\cos(x + 63{\cdot}9)^\circ$

We need to write the function in this form to analyse it easily.
We use the facts that $\cos x$ has: i) zeros at 90, 270,
ii) a max of 1 at 0, 360
iii) a min of −1 at 180.

$f(x) = 0 \Rightarrow x + 63{\cdot}9 = 90, 270 \Rightarrow x = 26{\cdot}1, 206{\cdot}1$

maximum value = $\sqrt{62}$, when $x + 63{\cdot}9 = 0, 360 \Rightarrow x = 296{\cdot}1$
minimum value = $-\sqrt{62}$, when $x + 63{\cdot}9 = 180 \Rightarrow x = 116{\cdot}1$

16·4 how to use the auxiliary angle to sketch the graph of $y = a\cos x + b\sin x$

Example Sketch the graph of $y = 2\sqrt{3}\cos x° - 5\sqrt{2}\sin x°$.

(Solution) This is the same function as in items 16·2 and 16·3.

We use all the information we have gathered there and transfer it to the graph, ensuring that everything is consistent.
The graph lies between $-\sqrt{62}$ and $+\sqrt{62}$, with min. t. pt. at $(116{\cdot}1, -\sqrt{62})$ and max. t. pt. at $(296{\cdot}1, \sqrt{62})$
It crosses the x-axis at $(26{\cdot}1, 0)$ and $(206{\cdot}1, 0)$
To find where it crosses the y-axis, put $x = 0$ in the original form, $2\sqrt{3}\cos x° - 5\sqrt{2}\sin x°$, rather than in $\sqrt{62}\cos(x + 63{\cdot}9)°$ to obtain $(0, 2\sqrt{3})$ hence:

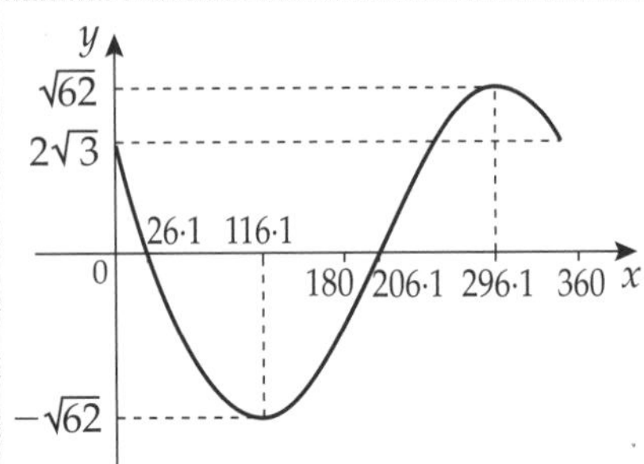

16·5 how to solve equations of the form $a\cos x + b\sin x = c$

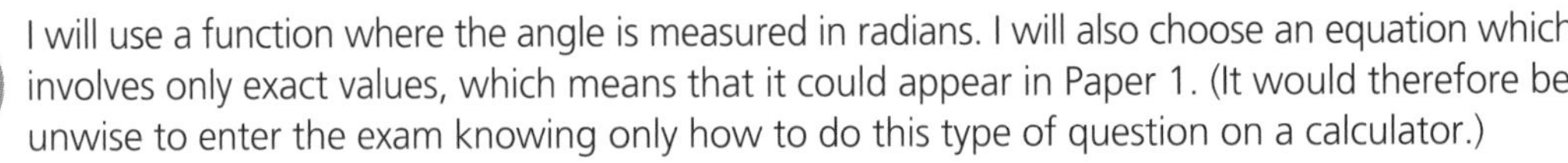

I will use a function where the angle is measured in radians. I will also choose an equation which involves only exact values, which means that it could appear in Paper 1. (It would therefore be unwise to enter the exam knowing only how to do this type of question on a calculator.)

Example Solve $\sin x - \sqrt{3}\cos x = \sqrt{2}$ for $0 \leqslant x \leqslant 2\pi$.

(Solution) Since a 'sin' term appears first,

let $\sin x - \sqrt{3}\cos x = k\sin(x - \alpha)$

$$= k(\sin x\cos\alpha - \cos x\sin\alpha)$$
$$= (k\cos\alpha)\sin x - (k\sin\alpha)\cos x$$

equating coefficients: $k\cos\alpha = 1$
$k\sin\alpha = \sqrt{3}$

$$\tan\alpha = \frac{k\sin\alpha}{k\cos\alpha} = \frac{\sqrt{3}}{1} = \sqrt{3} \Rightarrow$$

✓ sin	✓✓ All
tan	cos ✓

$$\Rightarrow \alpha = \frac{\pi}{3}$$

$k^2 = k^2\cos^2\alpha + k^2\sin^2\alpha = 1 + 3 = 4 \Rightarrow k = 2$ (since $k > 0$)

i.e. $\sin x - \sqrt{3}\cos x = 2\sin\left(x - \frac{\pi}{3}\right)$

so the equation becomes $2\sin\left(x - \frac{\pi}{3}\right) = \sqrt{2}$

Example continued ➢

Example continued

$$\Rightarrow \sin\left(x-\frac{\pi}{3}\right)=\frac{\sqrt{2}}{2}=\frac{1}{\sqrt{2}} \Rightarrow x-\frac{\pi}{3}=\frac{\pi}{4},\frac{3\pi}{4} \Rightarrow x=\frac{7\pi}{12},\frac{13\pi}{12}$$

(Alternatively
taking the 1 and the $\sqrt{3}$ as the shorter sides of a right angled triangle, the hypotenuse is 2, so divide through the equation by 2.

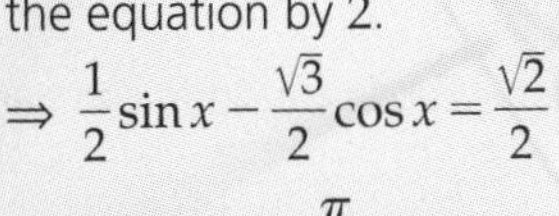

$$\Rightarrow \frac{1}{2}\sin x-\frac{\sqrt{3}}{2}\cos x=\frac{\sqrt{2}}{2}$$

$$\Rightarrow \sin x\times\cos\frac{\pi}{3}-\cos x\times\sin\frac{\pi}{3}=\frac{1}{\sqrt{2}}$$ (use the exact values judiciously)

$$\Rightarrow \sin\left(x-\frac{\pi}{3}\right)=\frac{1}{\sqrt{2}}$$ (use a compound angle formula)

$$\Rightarrow x-\frac{\pi}{3}=\frac{\pi}{4},\frac{3\pi}{4} \Rightarrow x=\frac{7\pi}{12},\frac{13\pi}{12}$$)

16·6 how to deal with $a\cos px+b\sin px$

An exercise of this complexity is only likely to appear near the end of Paper 2, if at all. (Here I restrict myself to what I have called method 1 in item 16·2.)

Example Find the maximum value of $f(x)=3\cos 2x°+4\sin 2x°+5$ and the value(s) of x between 0 and 360 for which it occurs.

(Solution) The function here is of the form $a\cos(A)+b\sin(A)$, so you would try to put it in a form like $k\cos(A-\alpha)$, and in this case A is $2x$.
Then, since the coefficients, 3 and 4, are both positive, it is easier to elect to express $(3\cos 2x°+4\sin 2x°)$ in the form $\cos(2x-\alpha)°$ or $\sin(2x+\alpha)°$. This will give us an acute auxiliary angle.
[If it had been $3\cos 2x°-4\sin 2x°$, I would have chosen $\cos(2x+\alpha)°$.]

Let $3\cos 2x°+4\sin 2x°=k\cos(2x-\alpha)°$
$\Rightarrow 3\cos 2x°+4\sin 2x°=k[\cos(2x)°\cos(\alpha)°+\sin(2x°)\sin(\alpha)°]$
$\Rightarrow 3\cos 2x°+4\sin 2x°=[k\cos(\alpha)°]\cos(2x)°+[k\sin(\alpha)°]\sin(2x)°$

equating coefficients: $k\cos\alpha°=3 \Rightarrow k^2\cos^2\alpha°=9$
$k\sin\alpha°=4 \Rightarrow k^2\sin^2\alpha°=16$
adding $k^2(\cos^2\alpha°+\sin^2\alpha°)=25$
$\Rightarrow k=5$ (since $k>0$)

dividing $\dfrac{k\sin\alpha}{k\cos\alpha}=\dfrac{4}{3} \Rightarrow \tan\alpha°=\dfrac{4}{3} \Rightarrow$

✓ sin	✓✓ All
tan	cos ✓

$\Rightarrow \alpha=53{\cdot}1$

$\Rightarrow f(x)=5\cos(2x-53{\cdot}1)°+5$
$\Rightarrow f_{max}=5\times1+5=10,$ when $2x-53{\cdot}1=0, 360$
$\Rightarrow 2x=53{\cdot}1, 413{\cdot}1 \Rightarrow x=26{\cdot}6, 206{\cdot}6$

$\Rightarrow$ the maximum value of f is 10 when $x=26{\cdot}6$ or $206{\cdot}6$

(Question 18·10 in Chapter 18 is also on this same topic as Examples 16·3, 16·4, 16·5 and 16·6.)

Chapter 17

TYPICAL SHORTER QUESTIONS

The questions given in this chapter are of a type similar to those found in the early to middle parts of a Higher Maths exam paper.

The examples already covered in the following items also come into this category, and are worth another look:

(3·7; 3·9; 3·13–3·16; 4·4; 4·8–4·11; 5·3–5·7; 5·9–5·11; 6·2; 6·4; 7·4; 7·7–8·1; 8·3–8·7; 9·3–9·8; 10·3–10·7; 11·2–11·3; 12·4–12·8; 13·6–13·12; 13·14; 14·1; 14·3; 15·3–15·5; 16·1–16·2.)

17·1 (The Straight Line)

Question and Answer

Show that the points A (1, 2), B (−2, −5) and C (5, −8) are three vertices of a square and find the co-ordinates of the fourth vertex.

Answer

(Three vertices of a square make up a right-angled isosceles triangle. You have been given the co-ordinates of these points. Make a sketch. You will then be looking at the same diagram as the person who wrote the question. The sketch suggests that AB = BC. Find their lengths.

y A (1, 2) O x B (−2, −5) C (5, −8)

$AB = \sqrt{(-2-1)^2 + (-5-2)^2} = \sqrt{9+49} = \sqrt{58}$

$BC = \sqrt{(-2-5)^2 + (-5-(-8))^2} = \sqrt{49+9} = \sqrt{58}$

$\Rightarrow AB = BC \Rightarrow \triangle ABC$ is isosceles

(To prove that $\triangle ABC$ is also right-angled we have a choice of methods.)

Method 1 (using the converse of the theorem of Pythagoras)

$AC^2 = (5-1)^2 + (-8-2)^2 = 16 + 100 = 116$

$AB^2 + BC^2 = 58 + 58 = 116 = AC^2$

$\Rightarrow \triangle ABC$ is right-angled at B (by the converse of the theorem of Pythagoras)

Method 2 (using gradients)

$m_{AB} = \frac{2-(-5)}{1-(-2)} = \frac{7}{3} \qquad m_{BC} = \frac{-8-(-5)}{5-(-2)} = \frac{-3}{7}$

$\Rightarrow m_{AB} \times m_{BC} = \frac{7}{3} \times \frac{-3}{7} = -1 \Rightarrow AB \perp BC$

Hence (by either method) $\triangle ABC$ is right-angled and isosceles i.e. A, B, C are vertices of a square.

Question and **Answer** continued ➢

Question and Answer continued

To find the fourth vertex, notice that A is 3 units along and 7 up from B, so D is 3 units along and 7 up from C hence D is $(5+3, \ -8+7)$ i.e. the fourth vertex is $(8, -1)$

(This could be written more formally using the vector $\begin{pmatrix}3\\7\end{pmatrix}$.)

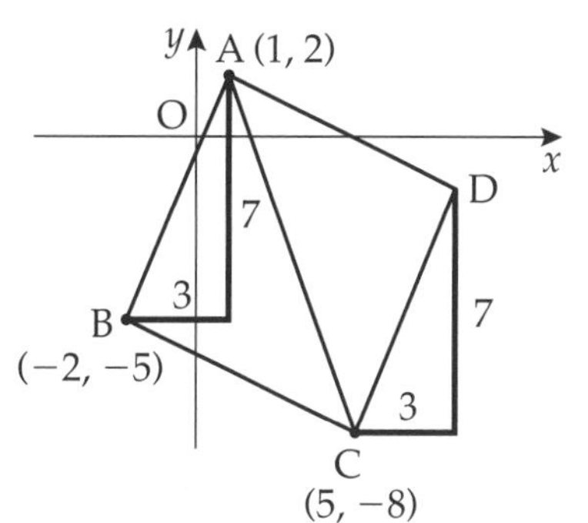

17·2 (The Straight Line)

Question and Answer

Find the equation of the line which passes through A, the point of intersection of the lines with equations $x+y=5$ and $y=x-1$, and which is perpendicular to the line with equation $4x+y=5$.

Answer

(A quick sketch here is again useful, but if it is going to take you too long to sketch these lines because you are not familiar enough with them, you would be better missing it out and spending the time directly on working which will score marks.)

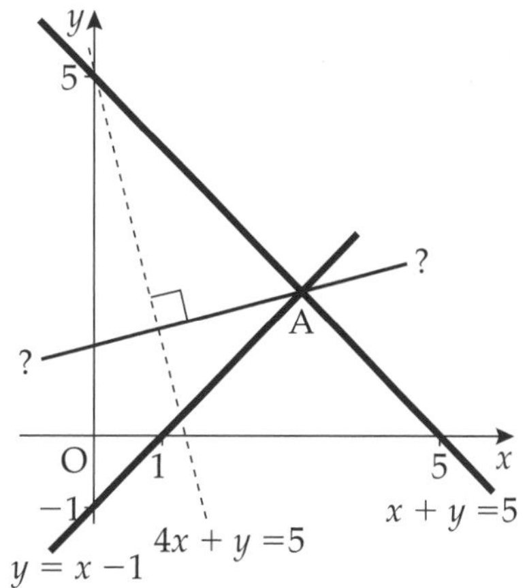

$x+y=5$ passes through 5 on both axes;
$y=x-1$ passes through $(0, -1)$ with gradient 1 so also passes through $(1, 0)$
$4x+y=5 \Rightarrow y=-4x+5$ which passes through $(0, 5)$ with gradient -4, [use $(1, 1)$ to add this line to the sketch]

(To find the equation of the new line, we need to know the co-ordinates of a point on it and its gradient, so that we can use $y-y_1=m(x-x_1)$.
The point A will lie on this new line. The gradient is found from that of $4x+y=5$.)

To find A, solve $x+y=5$ and $y=x-1$ simultaneously:
by substitution: $x+(x-1)=5$

$$2x=6 \Rightarrow x=3 \Rightarrow y=3-1=2$$

i.e. A is the point $(3, 2)$

Question and Answer continued ➤

Question and Answer continued

To find the gradient, consider the line $4x + y = 5$. Re-arrange as $y = -4x + 5$. This line has gradient -4, so the new line has gradient $\frac{1}{4}$ (using $m_1 \times m_2 = -1$)

Using $y - y_1 = m(x - x_1)$ gives $y - 2 = \frac{1}{4}(x - 3)$

$$\Rightarrow 4y - 8 = x - 3 \Rightarrow 4y = x + 5$$

17·3 (The Straight Line)

Question and Answer

Calculate correct to one decimal place the obtuse angle between the lines with equations $y = 2x + 1$ and $x + y = 1$

Answer

Both lines pass through (0, 1).
The required angle is $\alpha + \beta$, as shown.
α is the angle between $y = 2x + 1$ and Ox
$\Rightarrow \tan\alpha =$ the gradient of $y = 2x + 1$ (=2)
$\Rightarrow \alpha = 63{\cdot}4°$
γ is the angle between $y = -x + 1$ and Ox
$\Rightarrow \tan\gamma =$ the gradient of $y = -x + 1$ (= −1)
$\Rightarrow \gamma = 135°$
$\Rightarrow \beta = 180° - 135° = 45°$
hence $\alpha + \beta = 63{\cdot}4° + 45° = 108{\cdot}4°$

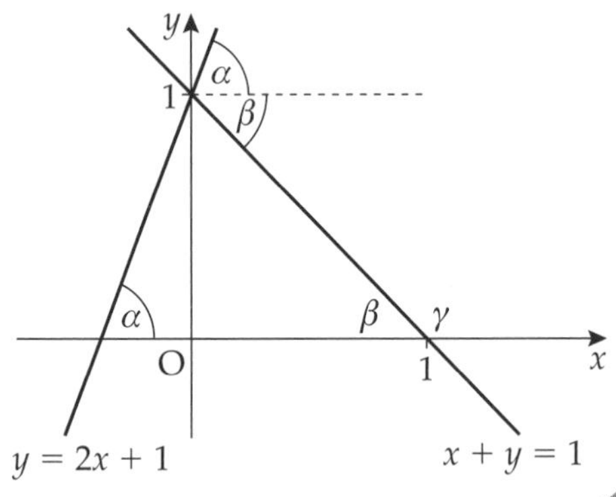

17·4 (The Straight Line)

Question and Answer

A is the point (8, 7), B(−2, −1), C(−3, 8) and D(1, t). Find the value of t for which AB is perpendicular to CD.

Answer

(We know that the perpendicular lines have gradients which multiply to give −1, so calculate the two relevant gradients.)

$$m_{AB} = \frac{7-(-1)}{8-(-2)} = \frac{8}{10} = \frac{4}{5} \qquad m_{CD} = \frac{8-t}{-3-1} = \frac{8-t}{-4}$$

Question and Answer continued ➢

Question and Answer continued

$$AB \perp CD \Rightarrow m_{AB} \times m_{CD} = -1$$
$$\Rightarrow \frac{4}{5} \times \frac{8-t}{-4} = -1$$
$$\Rightarrow \frac{8-t}{-5} = -1 \Rightarrow 8-t=5 \Rightarrow t=3$$

17·5 (Functions and Graphs)

Question and Answer

Two functions f and g are defined on $\mathbb{R}$, the set of all real numbers by

$f(x) = \frac{1}{4x-5},\ x \neq \frac{5}{4}$

$g(x) = x^2 + 1$

Find a) a formula for h where $h(x) = f(g(x))$
b) the largest possible domain for h.
c) a formula for $k(x) = f(f(x))$.

Answer

a) $h(x) = f(g(x)) = f(x^2+1) = \frac{1}{4(x^2+1)-5} = \frac{1}{4x^2-1}$

b) A fraction is not defined if the denominator is zero, so consider $4x^2 - 1 = 0$
i.e. $x = \pm\frac{1}{2}$, so the domain is $\left\{x \in \mathbb{R} \mid x \neq \pm\frac{1}{2}\right\}$

c) $k(x) = f(f(x)) = f\left(\frac{1}{4x-5}\right)$

$$= \frac{1}{4\left[\frac{1}{4x-5}\right]-5} = \frac{4x-5}{4-5(4x-5)} \quad \text{(multiply all 3 terms by } (4x-5))$$

$$= \frac{4x-5}{29-20x}$$

17·6 (Functions and Graphs)

Question and Answer

a) Express $x^2 - 10x + 21$ in the form $(x - a)^2 - b$.

b) Sketch the graph of the parabola with equation $y = f(x) = x^2 - 10x + 21$.

c) Sketch the graph of

i) $y = -f(x)$

ii) $y = f(-x)$

iii) $y = -f(-x)$

iv) $y = 42 - f(x)$.

Answer

a) $x^2 - 10x + 21 = x^2 - 10x + 25 - 25 + 21$ $\quad \left[\left(\frac{1}{2} \times (-10)\right)^2 = 25\text{, so add } 25 - 25\right]$

$= (x - 5)^2 - 4$

b) using part **a)**, y has a minimum value of -4 when $x = 5$.

$x = 0 \Rightarrow y = 21$

$\Rightarrow$ cuts the y-axis at $(0, 21)$

$y = 0 \Rightarrow (x - 5)^2 - 4 = 0$

$\Rightarrow x - 5 = \pm 2$

$\Rightarrow x = 3, 7$

$\Rightarrow$ cuts the x-axis at $(3, 0)$ and $(7, 0)$

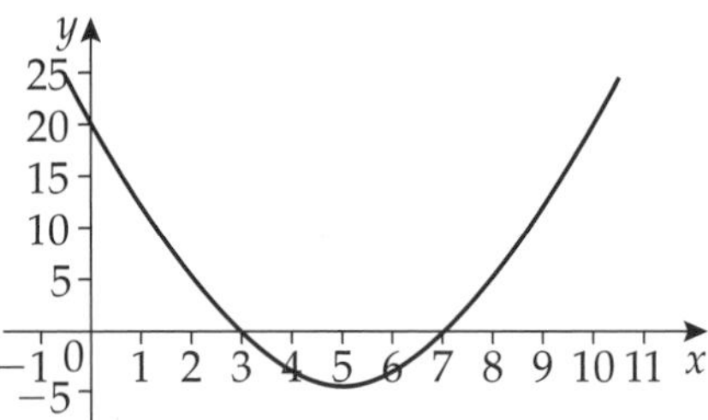

c) i) reflect in the x-axis

ii) reflect in the y-axis

iii) half turn about $(0, 0)$

iv) reflect in the x-axis then translate 42 units upwards parallel to the y-axis

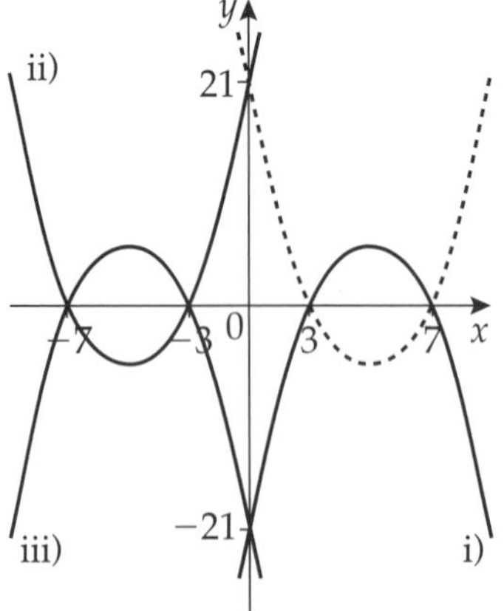

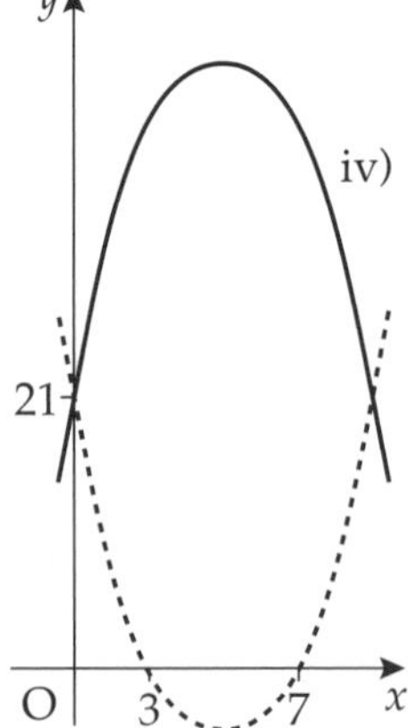

17·7 (Differentiation)

Question and Answer

This diagram shows the graph of $y = f(x)$. There is a minimum turning point at $(-a, -b)$ and a rising point of inflection at (c, d).

a) Sketch the graph of $y = f'(x)$.

b) If, on the given diagram, the gradient of f at the origin is 1, what additional information can you add to the sketch in your answer to part **a)**?

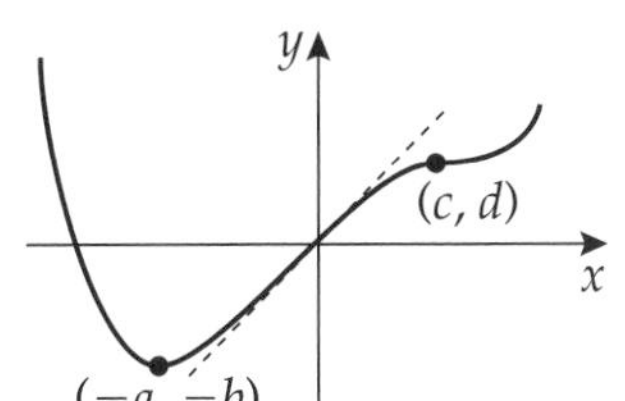

Answer

a) It is helpful to draw the graph of f' directly below the graph of f.

Basically you investigate three sub-domains: $x < -a$, $-a < x < c$, $x > c$.

f is stationary (i.e. $f' = 0$) when $x = -a$ and $x = c$ thus the graph of f' crosses the x-axis at $(-a, 0)$, and $(c, 0)$;

f is decreasing for $x < -a$, so the value of f' is negative i.e. the graph of f' will be below the x-axis;

f is increasing for $x > -a$, except at $x = c$, so the value of f' is positive i.e. the graph of f' is above the x-axis;

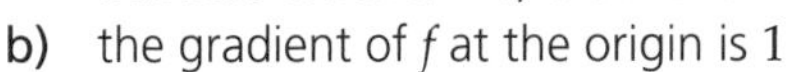

Since the graph of f' is positive on either side of $x = c$, but zero when $x = c$, the x-axis is a tangent at $x = c$.

b) the gradient of f at the origin is 1

$\Rightarrow f'(0) = 1 \Rightarrow$ when $x = 0, f' = 1$

$\Rightarrow$ the point $(0, 1)$ lies on the graph of f'

i.e. the graph crosses the y-axis at $(0, 1)$

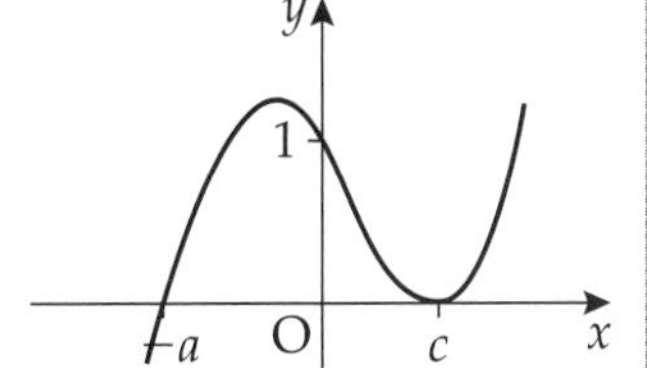

17·8 (Differentiation)

Question and Answer

For what values of x is the function $f(x) = 1 - 9x + 6x^2 - x^3$ decreasing?

Answer

The function f is decreasing where $f'(x) < 0$. (So find and factorise $f'(x)$)

now $f'(x) = -9 + 12x - 3x^2$

$\Rightarrow f'(x) = (-3)(x^2 - 4x + 3)$

$\Rightarrow f'(x) = (-3)(x - 1)(x - 3)$

f' has zeros at $x = 1$ and 3

f' has a maximum turning point because the coefficient of x^2 is negative, (-3)

Hence the sketch of f' is as shown

f' is negative when its graph is below the x-axis

i.e. for $\{x < -1\} \cup \{x > 3\}$

hence f is decreasing for $\{x < -1\} \cup \{x > 3\}$

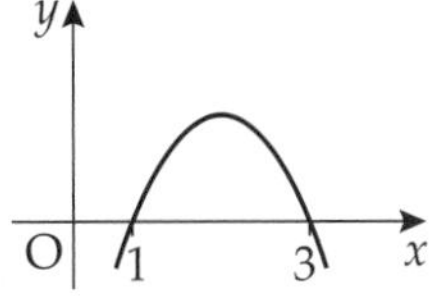

17·9 (Logarithms, Functions and Graphs)

Question and Answer

Find the values of p, q (and r in part **c)**) where:

a) $y = 2^{px} + q$

b) 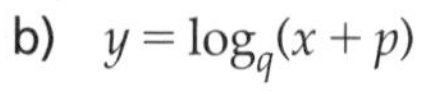
$y = \log_q(x + p)$

c) 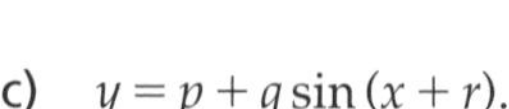
$y = p + q\sin(x + r)$.

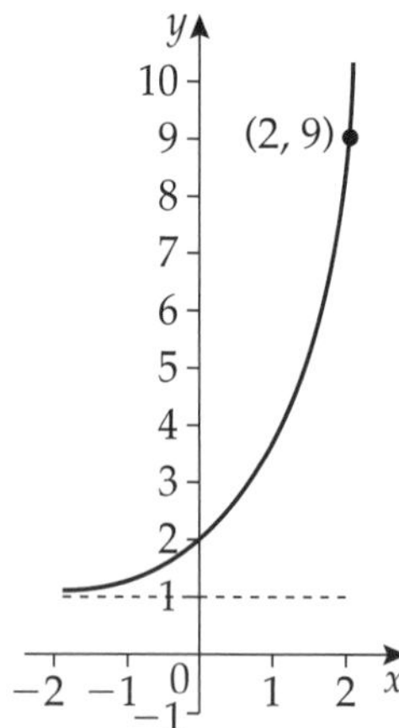

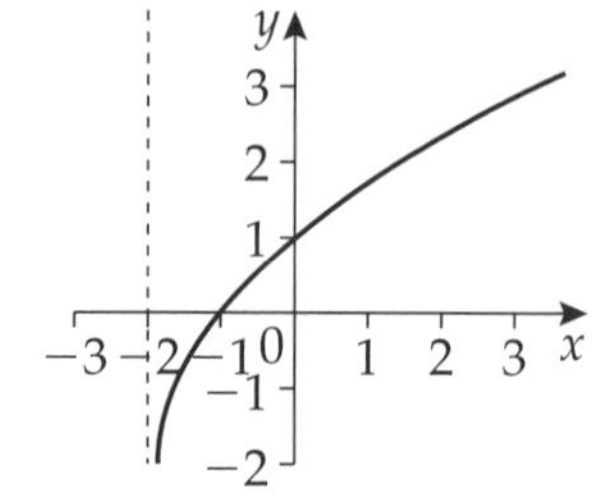

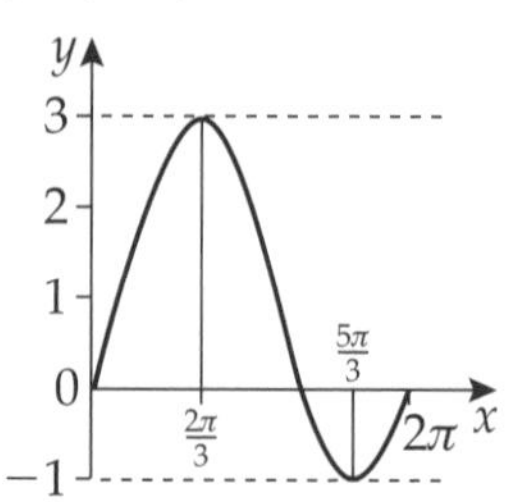

Answer

a) The asymptote for this exponential graph is $y = 1$.
So the graph of $y = 2^{px}$ has been translated up 1, i.e. $q = 1$.
The point (2, 9) was therefore previously the point (2, 8) on $y = 2^{px}$.
[or (2, 9) lies on the graph of $y = 2^{px} + 1$, so that $9 = 2^{2p} + 1 \Rightarrow 8 = 2^{2p}$]
$8 = 2^{2p} \Rightarrow 2^3 = 2^{2p} \Rightarrow 3 = 2p \Rightarrow p = \frac{3}{2}$.
Hence $p = \frac{3}{2}$, $q = 1$

b) The asymptote for this log graph is $x = -2$.
So the graph of $y = \log_q x$ has been moved 2 units left, i.e. $p = 2$.
The point (0, 1) was therefore previously the point (2, 1) on $y = \log_q x$.
[or (0, 1) lies on the graph of $y = \log_q(x + 2)$, so that $1 = \log_q(2)$]
$1 = \log_q 2 \Rightarrow q^1 = 2 \Rightarrow q = 2$
Hence $p = 2$, $q = 2$

c) This graph lies between $y = -1$ and $y = 3$. Half way between is $y = 1$.
So $y = q\sin(x + r)$ has been translated 1 unit up $\Rightarrow p = 1$
The graph goes 2 above and 2 below this middle line $\Rightarrow q = 2$
The graph of $y = 1 + 2\sin(x)$ has been moved right, so r is negative and acute.
We also know that (0, 0) lies on
$y = 1 + 2\sin(x + r) \Rightarrow 0 = 1 + 2\sin r$
i.e. $\sin r = -\frac{1}{2} \Rightarrow r = -\frac{\pi}{6}$
Hence $p = 1 \quad q = 2 \quad r = -\frac{\pi}{6}$

17·10 (Differentiation)

Question and Answer

If $y = 2x^2 + x - 1$, show that $x(1 + y') = 2(y + 1)$.

Answer

(Since we are asked to prove an identity involving y', differentiating might be a good start. But 'starting with the answer' is not a good idea. That is the line that we want to finish up with. We will need to start with one side (probably the left) of the identity in the answer and try to manipulate it to come out with the other side (the right).
Alternatively, simplify both sides independently till they are the same. If you take the answer as your first line of working and deduce that $1 = 1$ or $0 = 0$, then logically you have not really proved anything.)

Differentiating: $y' = 4x + 1$; also $y = 2x^2 + x - 1 \Rightarrow y + 1 = 2x^2 + x$

left hand side $= x(1 + y')$
$= x(1 + 4x + 1)$
$= x(4x + 2)$
$= 4x^2 + 2x$
$= 2(2x^2 + x)$
$= 2(y + 1)$
$=$ right hand side

(or from here, right hand side $= 2(y + 1)$
$= 2(2x^2 + x - 1 + 1)$
$= 2(2x^2 + x)$
$= 4x^2 + 2x$
hence left hand side $=$ right hand side

17·11 (Integration)

Question and Answer

a) Evaluate $\int_0^2 (x^2 - 4x + 7)dx$.

b) Make an appropriate sketch and shade the area represented by this definite integral.

Answer

a) $\int_0^2 (x^2 - 4x + 7)dx = \left[\frac{x^3}{3} - 2x^2 + 7x\right]_0^2 = \left[\frac{8}{3} - 8 + 14\right] - [0] = 6 + \frac{8}{3} = \frac{26}{3}$

b) We have to sketch the graph of '$y =$ the integrand' and shade under the curve from 0 to 2.
The curve is $y = x^2 - 4x + 7$
$= x^2 - 4x + 4 - 4 + 7$
$= (x - 2)^2 + 3$
This has min. t. pt. (2, 3)
and passes through (0, 7). Hence:

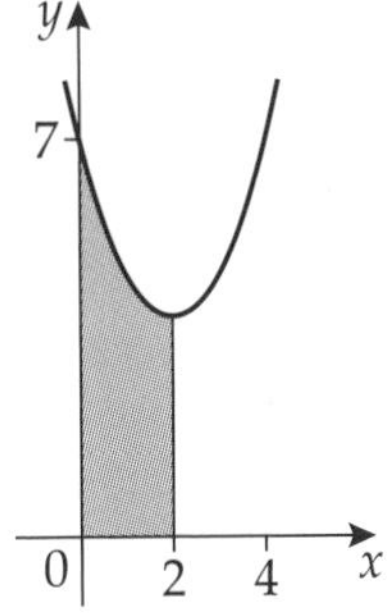

17·12 (Identities and Radians)

Question and Answer

Solve $\cos x < -\frac{1}{2}$, for $0 < x < 2\pi$.

Answer

Try this question for yourself before reading on.

'That's not fair. We've never done anything like this in class before. It's not listed in your itemisation of the syllabus,' I hear you cry. You could possibly argue that way, but you know enough Mathematics to answer the question. The setters are allowed to ask you questions which make you think. The solution is relatively straightforward. The first line of attack when you are stuck should always be 'can I draw a sketch?'

So let us start with the graph of $y = \cos x$.

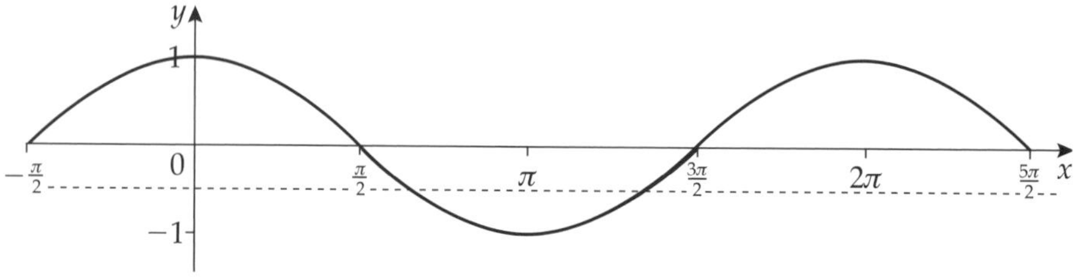

Let us add the line $y = -\frac{1}{2}$.

We want to read off where the graph is below the line $y = -\frac{1}{2}$.

When is $\cos x = -\frac{1}{2}$? When $x = \pi - \frac{\pi}{3}$ and when $x = \pi + \frac{\pi}{3}$ i.e. $\frac{2\pi}{3}, \frac{4\pi}{3}$

Hence $\cos x < -\frac{1}{2}$ for $\frac{2\pi}{3} < x < \frac{4\pi}{3}$

17·13 (The Circle)

Question and Answer

Find the length of the tangent from P(21, −13) to the circle with equation $x^2 + y^2 - 2x - 4y - 44 = 0$

Answer

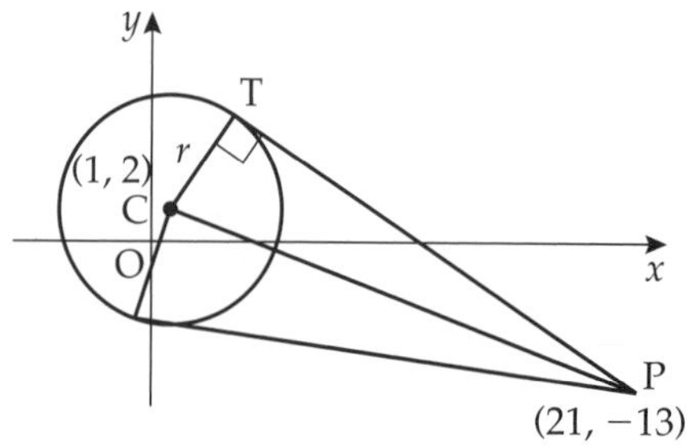

(Not having the sketch given in the question makes it much harder. The key to solving this problem lies in deciding to draw your own diagram. Remember that the tangent is perpendicular to the radius and mark the right angles.
Once you look at the sketch, you should realise immediately that this is merely a Pythagoras calculation.)

Question and **Answer** continued ➢

Question and Answer continued

Centre C is (1, 2) and radius $r = \sqrt{1 + 4 + 44} = 7$

$PC = \sqrt{20^2 + 15^2} = \sqrt{625} = 25$ (by the distance formula, item **3·1**)

$\Rightarrow PT^2 = PC^2 - CT^2 = 625 - 49 = 576$ (by the theorem of Pythagoras)

$\Rightarrow PT = \sqrt{576} = 24$

$\Rightarrow$ the length of tangent $= 24$

17·14 (Vectors)

Question and Answer

V, ABCD is a right-rectangular pyramid.

$\overrightarrow{AB} = \mathbf{p} \quad \overrightarrow{AD} = \mathbf{q} \quad \overrightarrow{AV} = \mathbf{r}$

Express $\overrightarrow{CV}$ in terms of $\mathbf{p}$, $\mathbf{q}$ and $\mathbf{r}$.

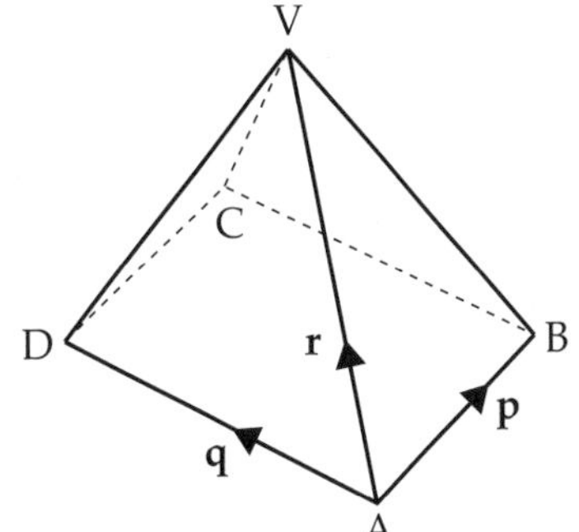

Answer

(Make your own skeleton diagram of the pyramid, and add the line AC.)

We need to think of a pathway which takes us from C to V, using the 'paths' $\mathbf{p}$, $\mathbf{q}$ and $\mathbf{r}$.

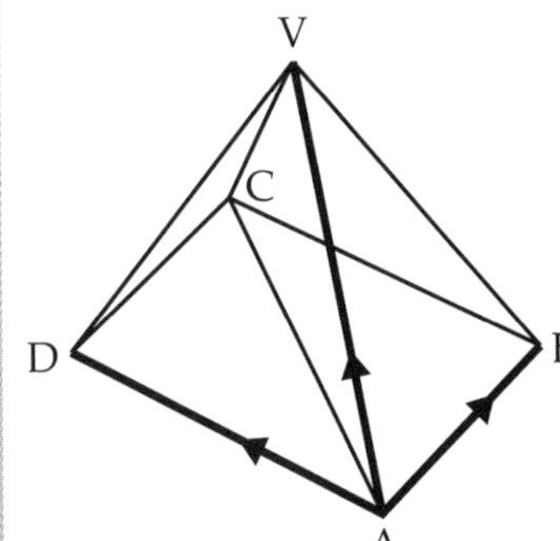

$$\begin{aligned}
\overrightarrow{CV} &= \overrightarrow{CA} + \overrightarrow{AV} \\
&= \overrightarrow{AV} - \overrightarrow{AC} \\
&= \mathbf{r} - (\overrightarrow{AB} + \overrightarrow{BC}) \\
&= \mathbf{r} - (\mathbf{p} + \overrightarrow{AD}) \\
&= \mathbf{r} - \mathbf{p} - \mathbf{q}
\end{aligned}$$

17·15 (Identities, Compound and Multiple Angles)

Question and Answer

If P is an acute angle such that $\sin P = \frac{1}{\sqrt{5}}$, find the exact value of a) $\cos P$ b) $\sin 2P$.

Question and Answer continued ➢

Question and Answer continued

Answer

a) Either

draw a right-angled triangle with an angle P in it, and mark the appropriate sides as 1 and $\sqrt{5}$. Then calculate the third side (by Pythagoras' Theorem)

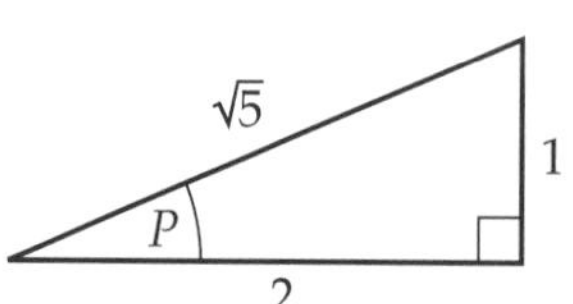

so $\cos P = \frac{2}{\sqrt{5}}$

or

use the identity $\cos^2 x + \sin^2 x = 1$

$\Rightarrow \cos^2 P = 1 - \sin^2 P$

$\Rightarrow \cos^2 P = 1 - \left(\frac{1}{\sqrt{5}}\right)^2$

$\Rightarrow \cos^2 P = 1 - \frac{1}{5} = \frac{4}{5}$

$\Rightarrow \cos P = \pm\frac{2}{\sqrt{5}}$

since P is acute $\cos P = \frac{2}{\sqrt{5}}$

b) $\sin 2P = 2\sin P\cos P = 2 \times \left(\frac{1}{\sqrt{5}}\right) \times \left(\frac{2}{\sqrt{5}}\right) = \frac{4}{5}$

17·16 (The Circle)

Question and Answer

Find the points of intersection of the line with equation $2x - y + 5 = 0$ and the circle with equation $x^2 + y^2 = 50$.

Answer

carefully re-arrange the linear equation:	$y = 2x + 5$
substitute in the circle equation:	$x^2 + (2x+5)^2 = 50$
expand:	$x^2 + 4x^2 + 20x + 25 - 50 = 0$
simplify:	$5x^2 + 20x - 25 = 5(x^2 + 4x - 5) = 0$
factorise:	$(x-1)(x+5) = 0$
solve for x:	$x = 1, -5$
use $y = 2x + 5$ to find y:	$y = 7, -5$

Hence the points of intersection are $(1, 7)$, $(-5, -5)$

17·17 (Identities and Radians)

Question and Answer

The function $Q(\theta)$ is defined by $Q(\theta) = 5\sin\left(2\theta - \frac{\pi}{3}\right)$.

Write down the maximum value of Q, and find the values of θ $(0 < \theta < 2\pi)$ for which this maximum value occurs.

Answer

The maximum value of $\sin$ (anything) is 1.

$\Rightarrow$ The maximum value of $5\sin$ (anything) is 5. Hence 5

The maximum value of $\sin(x)$ occurs when x is $\frac{\pi}{2}$.

$\Rightarrow$ The maximum value of Q occurs when $\left(2\theta - \frac{\pi}{3}\right)$ is $\frac{\pi}{2}, \frac{5\pi}{2}$

Remember to include the second angle because of the 2 in front of θ.

$$\Rightarrow 2\theta = \frac{\pi}{2} + \frac{\pi}{3}, \frac{5\pi}{2} + \frac{\pi}{3}$$

$$= \frac{5\pi}{6}, \frac{17\pi}{6}$$

$$\Rightarrow \quad \theta = \frac{5\pi}{12}, \frac{17\pi}{12}$$

17·18 (Differentiation)

Question and Answer

The straight line in this graph has equation $y = f(x)$.
Draw the graph of the derived function y'.

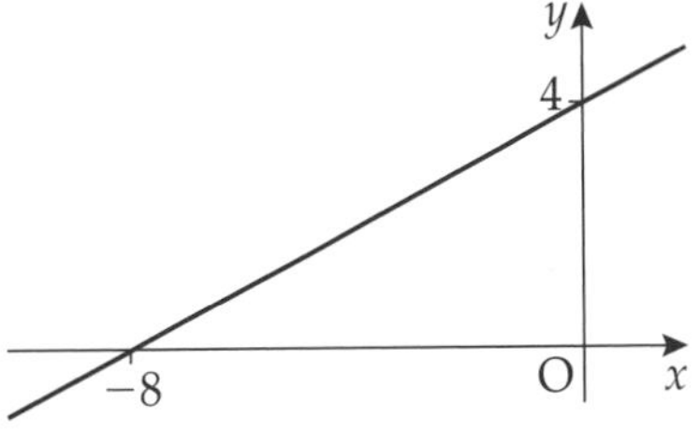

Answer

The derived function measures the gradient of a function.

The gradient of this line is constant, namely $\frac{1}{2}$.

So the graph has equation $y' = \frac{1}{2}$.

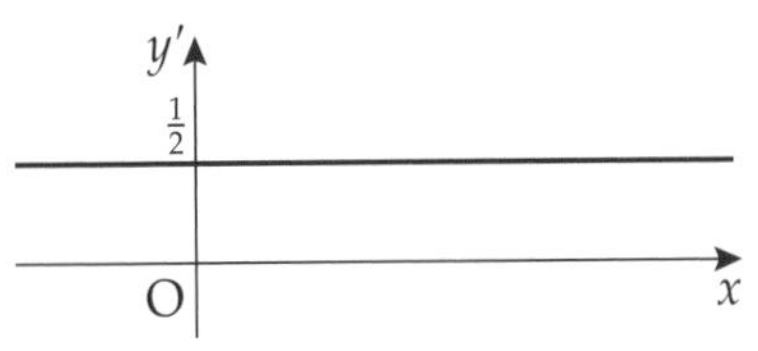

17·19 (The Circle)

Question and Answer

PQ and RS are the tangents from P (0, 13) to the circle with centre R, and equation $x^2 + y^2 = 25$.

Calculate the area of the kite PQRS.

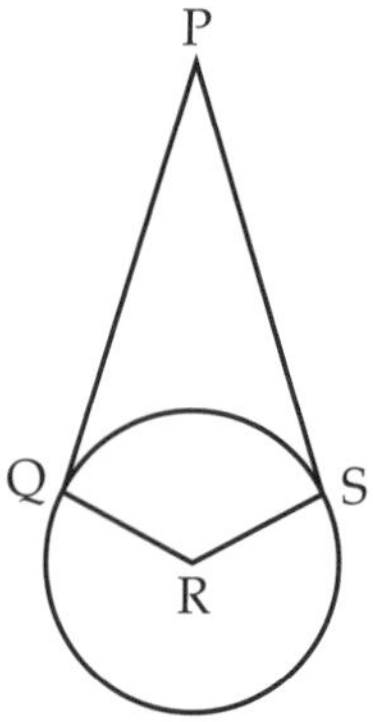

Answer

(Remember that the tangent is perpendicular to the radius, and PR is an axis of symmetry.)

$x^2 + y^2 = 25 \Rightarrow$ R is (0, 0) and QR = 5 (= RS)

R is (0, 0) and P is (0, 13) $\Rightarrow$ PR = 13

So $\triangle$PRS is a '5, 12, 13'$\triangle$ $\Rightarrow$ PS = 12

Area of PQRS = 2 × area of $\triangle$PRS $= 2 \times \frac{1}{2} \times 5 \times 12 = 60$ units²

17·20 (Recurrence relations)

Question and Answer

Two sequences are defined by the recurrence relations

$$u_{n+1} = 0{\cdot}4u_n + p, \quad u_0 = 10$$
$$v_{n+1} = 0{\cdot}6v_n + p^2, \quad v_0 = 10$$

If both sequences have the same non-zero limit, find the exact value of p.

Answer

$-1 < 0{\cdot}4 < 1$, and $-1 < 0{\cdot}6 < 1$, confirm that each sequence has a limit.

Let this common limit be L; use the formula $L = \frac{b}{1-a}$

then from $\{u_n\}$; $L = \frac{p}{1-0{\cdot}4}$, and from $\{v_n\}$; $L = \frac{p^2}{1-0{\cdot}6}$

Hence $\frac{p}{0{\cdot}6} = \frac{p^2}{0{\cdot}4}$

$\Rightarrow 0{\cdot}6p^2 = 0{\cdot}4p \Rightarrow \frac{p^2}{p} = \frac{0{\cdot}4}{0{\cdot}6} = \frac{2}{3}$ i.e. $p = \frac{2}{3}$

Do not write $\frac{2}{3}$ as 0·6, or 0·66 or 0·667.

Only the vulgar fraction (or $0{\cdot}\dot{6}$) is exact.

17·21 (The Straight Line)

Question and Answer

The line l makes an angle of $\frac{\pi}{3}$ radians with the y-axis as shown.

Find the exact value of the gradient of l.

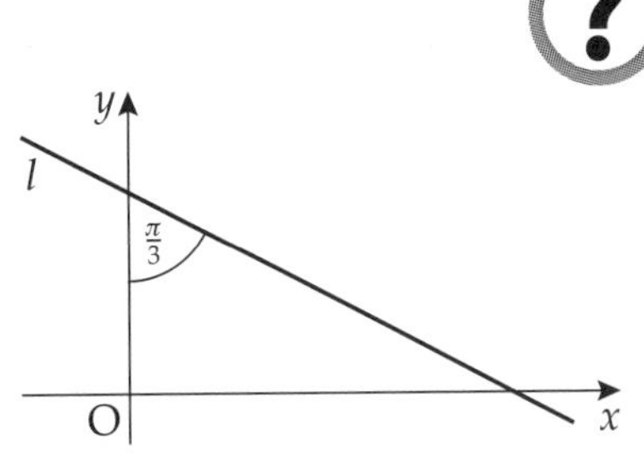

Answer

The gradient of $l = \tan b$.

$a = \frac{\pi}{2} - \frac{\pi}{3} = \frac{\pi}{6} \Rightarrow b = \pi - \frac{\pi}{6} = \frac{5\pi}{6}$

hence the gradient of $l = \tan\left(\frac{5\pi}{6}\right) = -\frac{1}{\sqrt{3}}$

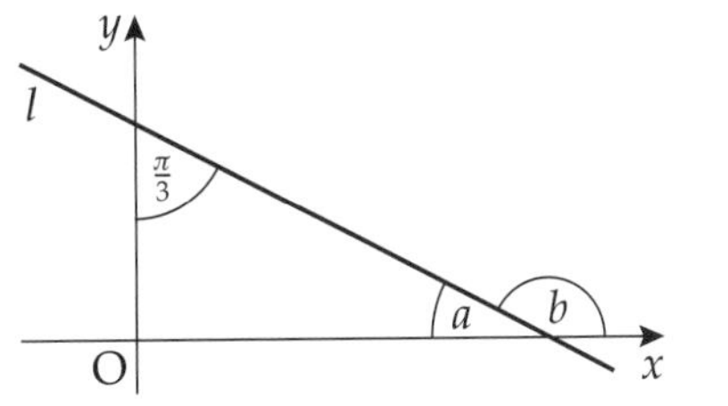

17·22 (Vectors, The Circle)

Question and Answer

A, B, C are the points $(1, -1, -2)$, $(3, 3, 2)$ and $(4, 2, t)$ respectively.

C lies on a circle which has AB as a diameter.

Find all the possible values of t, if any.

Answer

(There are lots of circles in three dimensions with AB as diameter. We do not know how to find their equations, so that cannot be the expected solution.

There are two obvious methods of solution:

1) find the centre P, and hence the radius; use the distance formula for CP to find t.
2) $\hat{C}$ must be an angle in a semi-circle, so $\hat{C}$ is a right angle; hence $CA \perp CB$, so the vectors $\overrightarrow{CA}$ and $\overrightarrow{CB}$ must have a scalar product of zero.)

Method 1

P is the mid point of AB $\Rightarrow$ P is $(2, 1, 0)$

the radius = PA $\Rightarrow r = \sqrt{(2-1)^2 + (1-(-1))^2 + (0-(-2))^2} = \sqrt{9} = 3$

$\Rightarrow CP = 3$

but $CP = \sqrt{(4-2)^2 + (2-1)^2 + (t-0)^2} = \sqrt{5 + t^2}$

so $\sqrt{5 + t^2} = 3 \Rightarrow 5 + t^2 = 9 \Rightarrow t = \pm 2$

Question and **Answer** continued ➢

Question and Answer continued

Method 2

$\overrightarrow{AC} = \mathbf{c} - \mathbf{a} = \begin{pmatrix} 3 \\ 3 \\ t+2 \end{pmatrix}$ $\quad \overrightarrow{BC} = \mathbf{c} - \mathbf{b} = \begin{pmatrix} 1 \\ -1 \\ t-2 \end{pmatrix}$

$\overrightarrow{AC}.\overrightarrow{BC} = 0 \Rightarrow (3 \times 1) + (3 \times (-1)) + (t+2)(t-2) = 0$

$\Rightarrow t = \pm 2$ (as before)

17·23 (Quadratic Theory)

Question and Answer

A curve has equation $y = 3x^3 + 4x^2 + 5x - 6$.
Prove that this curve has no stationary points.

Answer

The derived function is zero at every stationary point, so the question becomes a matter of showing that the derivative cannot be zero.

$y' = 9x^2 + 8x + 5 = 0$ for stationary points

You must recognise this as a quadratic equation. The question does not tell you. If it has no solutions, the discriminant will be negative.

Note: calculus and algebra are mixed together in the same question here, and again in 17·24.

$\triangle = b^2 - 4ac = 8^2 - 4 \times 9 \times 5 = 64 - 180 < 0 \Rightarrow$ no real roots of $y' = 0$

$\Rightarrow$ no stationary points on y

17·24 (Differentiation)

Question and Answer

The point P$(-1, p)$ lies on the graph with equation $y = 3x^3 - 4x^2 - 9x + 2$.

a) Find the value of p.
b) Prove that this graph is increasing at P.

Answer

a) When $x = -1$, $y = p \Rightarrow p = -3 - 4 + 9 + 2 = 4$ i.e. $p = 4$

b) $y' = 9x^2 - 8x - 9 \Rightarrow y'(-1) = 9 + 8 - 9 = 8(>0) \Rightarrow y$ is increasing

Notice that parts **a)** and **b)** are independent. Try **b)** even if you cannot do **a)**

17·25 **(Quadratic Theory)**

Question and Answer

Find the value(s) of k for which the equation $(x+2)(x+k) = -9$ has equal roots.

Answer

This is a standard sort of question. I include it here to remind you that you must re-arrange it in standard form before you try to calculate the discriminant.

$$\begin{aligned}(x+2)(x+k) = -9 &\Rightarrow x^2 + 2x + kx + 2k + 9 = 0\\ &\Rightarrow x^2 + (2+k)x + (2k+9) = 0\end{aligned}$$

$$\begin{aligned}\triangle = b^2 - 4ac &= (k+2)^2 - 4 \times 1 \times (2k+9)\\ &= k^2 + 4k + 4 - 8k - 36\\ &= k^2 - 4k - 32\\ &= (k+4)(k-8) = 0 \text{ for equal roots} \Rightarrow k = -4, 8\end{aligned}$$

17·26 **(Further Calculus)**

Question and Answer

Evaluate $f'(2)$ where $f(x) = 3(2x-5)^4$.

Answer

$f(x) = 3(2x-5)^4 \Rightarrow f'(x) = 3 \times 4 \times (2x-5)^3 \times 2$

The last '2' is the derivative of $(2x-5)$.

$$\Rightarrow f'(2) = 12 \times (-1)^3 \times 2 = -24$$

This is a standard question on unit 3 calculus. It is considered to be more difficult than grade C. I include it to remind you of the most common error, namely omitting the derivative of the bracket as part of the chain rule. Have another look at item **14·2a**.

17·27 **(Further Calculus)**

Question and Answer

Evaluate $g'\left(\frac{7\pi}{6}\right)$ where $g(x) = 3(1 + 2\sin x)^2$.

Answer

(Additional errors occur with the substitution in this type of question.)

$$\begin{aligned}g(x) = 3(1+2\sin x)^2 &\Rightarrow g'(x) = 3 \times 2 \times (1 + 2\sin x) \times (2\cos x)\\ &\Rightarrow g'\left(\frac{7\pi}{6}\right) = 6\left[1 + 2\sin\left(\frac{7\pi}{6}\right)\right]\left[2\cos\left(\frac{7\pi}{6}\right)\right]\\ &= 6 \times \left[1 + 2\left(-\frac{1}{2}\right)\right] \times \left[2 \times \left(-\frac{\sqrt{3}}{2}\right)\right] = 6 \times 0 \times (-\sqrt{3}) = 0\end{aligned}$$

Chapter 18

TYPICAL LONGER QUESTIONS

The questions given in this chapter are of a type similar to those found in the middle to later parts of a Higher Maths exam paper.

The examples already covered in the following items also come into this category, and are worth another look:

(9·8; 13·13; 14·2; 15·7; 16·3–16·6.)

18·1 (The Straight Line)

Question and Answer

Triangle ABC has vertices A(4, 14), B(−5, −4) and C(13, 2).

Find
a) the equation of the altitude BD
b) the equation of l, the perpendicular bisector of BC
c) the co-ordinates of E, the point of intersection of l and BD
d) the co-ordinates of F, the point where the lines CE and AB meet.

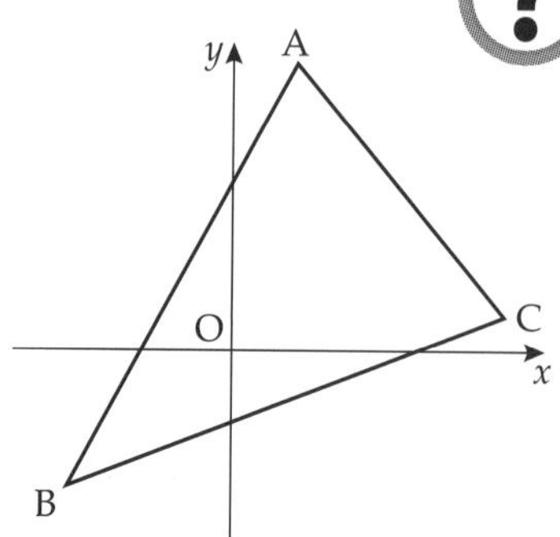

Answer

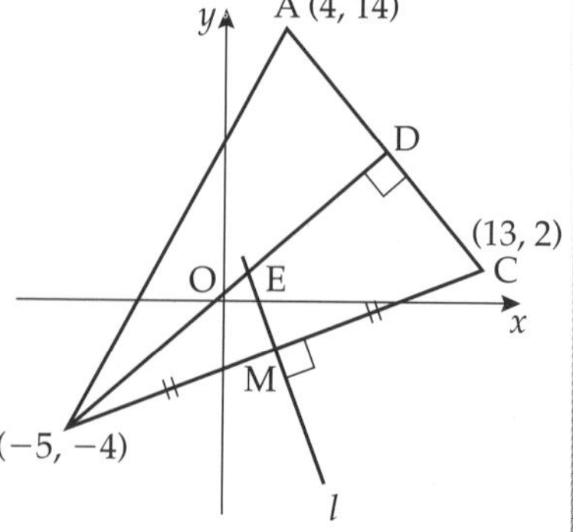

a) $m_{AC} = \dfrac{14-2}{4-13} = \dfrac{12}{-9} = -\dfrac{4}{3}$

$\Rightarrow m_{BD} = \dfrac{3}{4}$ (using $m_1 \times m_2 = -1$)

$\Rightarrow y - (-4) = \dfrac{3}{4}(x+5)$ (using $y - y_1 = m(x - x_1)$)

$\Rightarrow 4y + 16 = 3x + 15$

$\Rightarrow 4y = 3x - 1$

b) M is $(4, -1)$, $m_{BC} = \dfrac{2-(-4)}{13-(-5)} = \dfrac{6}{18} = \dfrac{1}{3} \Rightarrow m_l = -3$ (using $m_1 \times m_2 = -1$)

$\Rightarrow l$ has equation $y - (-1) = -3(x - 4)$ (using $y - y_1 = m(x - x_1)$)

$\Rightarrow y + 1 = -3x + 12 \Rightarrow 3x + y = 11$

Question and **Answer** continued ➢

Question and Answer continued

c) solve the equations simultaneously:
l has equation $y = 11 - 3x$: substituting in to $4y = 3x - 1$ gives $4(11 - 3x) = 3x - 1$
$\Rightarrow 44 - 12x = 3x - 1$
$\Rightarrow 45 = 15x \Rightarrow x = 3 \Rightarrow y = 11 - 3 \times 3 = 2$
$\Rightarrow$ E is the point $(3, 2)$

d) (find the equations of AB and CE and solve them simultaneously):

$$m_{AB} = \frac{14-(-4)}{4-(-5)} = \frac{18}{9} = 2; \quad \text{using } y - y_1 = m(x - x_1) \text{ gives } y - (-4) = 2(x - (-5))$$
$$\Rightarrow y + 4 = 2x + 10$$
$$\Rightarrow y = 2x + 6$$

C is $(13, 2)$ and $E(3, 2)$ $\Rightarrow$ CE has equation $y = 2$; (easy! remember item **3·12**)
substitute $y = 2$ in $y = 2x + 6 \Rightarrow 2 = 2x + 6 \Rightarrow x = -2 \Rightarrow$ F
is the point $(-2, 2)$

18·2 **(Differentiation)**

Question and Answer

The diagram shows the end wall of a room in a castle where the roof is barrel-vaulted.

It is intended to install a rectangular window in this wall, with one corner of the window at the bottom left hand corner where the floor meets the vertical wall.

If the floor is taken as the x-axis and the vertical wall as the y-axis, then the underside of the roof can be represented by the parabola $y = 15 - 2x - x^2$. (All measurements are in metres.)

Show that the area of the window cannot exceed $15\,\text{m}^2$.

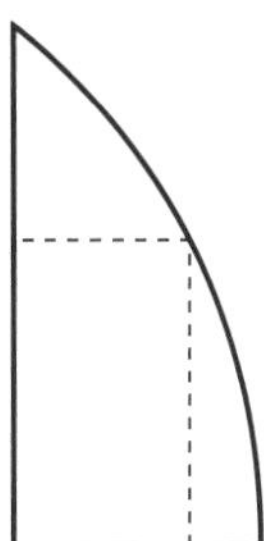

Answer

(This question is unusual in that you are not asked to prove a formula in 'part a' for the area, the quantity in which we are interested. The modelling only takes two lines, so we should manage it without assistance. Nor does the question ask 'Find the maximum area of the window', but that essentially is what we must do. Let us make our own mathematical diagram for this situation).

***Question** and **Answer** continued* ➢

Question and Answer continued

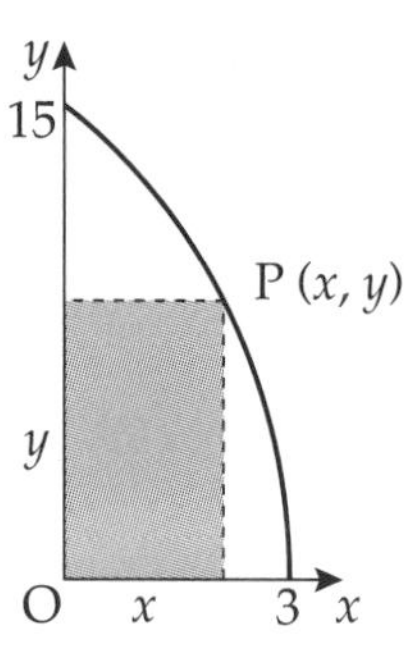

for the parabola, when $x = 0$, $y = 15$ hence (0, 15)

$y = 0 \Rightarrow 15 - 2x - x^2 = 0$

$\Rightarrow (5 + x)(3 - x) = 0$

$\Rightarrow x = -5, 3$ hence (3, 0)

Let the area of the window be $A\,\text{m}^2$

then $A = x \times y$

$= x \times (15 - 2x - x^2)$

$= 15x - 2x^2 - x^3$

so we need the greatest value of A in the interval $0 \leqslant x \leqslant 3$

$A'(x) = 15 - 4x - 3x^2$

$= (5 - 3x)(3 + x) = 0$ for stationary points $\Rightarrow x = -3, \frac{5}{3}$

$x = -3$ is outwith the domain we require, so $x = \frac{5}{3}$

Either

x	$(\frac{5}{3})^-$	$\frac{5}{3}$	$(\frac{5}{3})^+$
$(5-3x)$	+	0	−
$(3+x)$	+	+	+
$(5-3x)(3+x) = A'$	+	0	−
(tangent)	↗	→	↘

or

$A'' = -4 - 6x$

$\Rightarrow A'' = -4 - 10 (< 0)$

$\Rightarrow$ maximum

When $x = \frac{5}{3}$,

$A = 15x - 2x^2 - x^3 = 15\left(\frac{5}{3}\right) - 2\left(\frac{25}{9}\right) - \left(\frac{125}{27}\right) = 25 - \frac{150}{27} - \frac{125}{27} = 25 - 10\frac{5}{27} = 14\frac{22}{27}$

the maximum value of A is less than 15,
so the area of the window cannot exceed $15\,\text{m}^2$.

18·3 (Recurrence Relations)

Question and Answer

'Molarity' is a new (fictitious) genetically-modified product designed to prevent moles from attacking lawns. It is recommended that for a $100\,\text{m}^2$ lawn, 5 g of 'Molarity' should be dissolved in a gallon of water and sprayed evenly over the lawn once a week. Exceeding this does burns the grass. Worm activity reduces the amount of 'Molarity' present in the soil by 30% per week.

Question and Answer continued ➢

Question and Answer

a) 'Molarity' only becomes effective against moles once there is $8\,\text{g}$ of it per $100\,\text{m}^2$ continuously in the soil. After how many applications of 'Molarity' will a lawn become effectively protected against moles?

b) If the concentration of 'Molarity' in the soil exceeds $16\,\text{g}$ per $100\,\text{m}^2$, the worms will die. Is it safe to use Molarity to protect the lawn indefinitely?

Answer

a) (Notice that a 30% loss means 70% is retained. I recommend drawing a saw tooth diagram to envisage how the amount of 'Molarity' varies initially.)

5, after the first dose, drops to $0{\cdot}7 \times 5\,\text{g}$, i.e. $3{\cdot}5\,\text{g}$ then rises to $8{\cdot}5\,\text{g}$, drops to $0{\cdot}7 \times 8{\cdot}5\,\text{g}$, i.e. $5{\cdot}95\,\text{g}$, etc.

From this diagram, it is clear that $8\,\text{g}$ is only continuously achieved after 4 applications

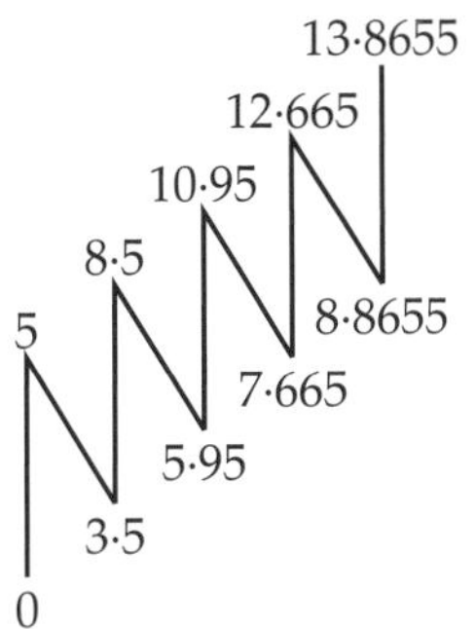

b) (Set up a recurrence relation to model this situation, use an algebraic method of establishing the limit, and interpret your answer.

Draw a general part of the saw tooth diagram to obtain the recurrence relation.)

This diagram is enough to justify that the appropriate recurrence relation is $u_{n+1} = 0{\cdot}7u_n + 5$

u_n $(0{\cdot}7 \times u_n) + 5$ $0{\cdot}7 \times u_n$

(Remember to justify that a limit exists.)

$-1 < 0{\cdot}7 < 1 \Rightarrow$ there is a limit, l say,

given by $l = 0{\cdot}7l + 5$*

$\Rightarrow\ 0{\cdot}3l = 5$

$\Rightarrow\ l = \dfrac{5}{0{\cdot}3} = \dfrac{50}{3} = 16\dfrac{2}{3}$

In other words, if the lawn is treated indefinitely with 'Molarity', the worms will die.

* or alternatively using $l = \dfrac{b}{1-a}$, $l = \dfrac{5}{1-0{\cdot}7} = \dfrac{5}{0{\cdot}3}$ as above

(If you used the recurrence relation quoted in part **b)** for part **a)**, it would give you the values at the top of the saw tooth diagram. Each of those at the bottom could easily be obtained by subtracting 5 from the number directly above it.)

18·4 **(Differentiation)**

Question and Answer

The Muphson Stage, carrying gold bullion to pay the cavalry, is thundering south on the Dry Bones Trail hotly pursued by the Bar-L outlaws.

In order to head off the stage at the Tiel Pass, shown on the accompanying map, half of the outlaws take the Indian Track, a single pathway through the woods which leaves the Dry Bones Trail tangentially just before Snake Bend.

Relative to suitable co-ordinate axes, the equation of the Dry Bones Trail is $y = x^3 - 2x^2 + x - 1$, and the Indian Track is the tangent to this curve at the point (2, 1).

a) Find the co-ordinates of the point where the outlaws on the track regain the trail.
b) Calculate the area enclosed between the Dry Bones Trail and the Indian Track.

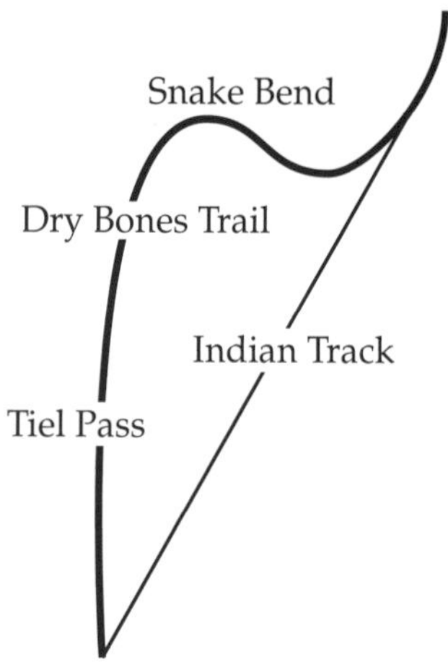

Answer

a) (This part falls into two basic calculations; find the equation of the tangent to a curve (see item **5·5**); find the intersection of a line and a curve (see items **8·7** and **12·7**))

$y = x^3 - 2x^2 + x - 1 \Rightarrow y' = 3x^2 - 4x + 1;$

the gradient at the point (2, 1) $= y'(2) = 12 - 8 + 1 = 5;$
so the equation of the track is $y - 1 = 5(x - 2)$ i.e. $y = 5x - 9$

To find where this line intersects the cubic curve, substitute the linear equation into the cubic equation, i.e. $5x - 9 = x^3 - 2x^2 + x - 1$
$\Rightarrow x^3 - 2x^2 - 4x + 8 = 0$

(although we need to use the remainder theorem to factorise this cubic expression, we already know that $x = 2$ is a double root of this equation, as the line is a tangent at the point where $x = 2$. Proceed as in item **9·5**.)

so 2	1	−2	−4	8
		2	0	−8
	1	0	−4	0

$\Rightarrow (x - 2)(x^2 - 4) = 0$
$\Rightarrow (x - 2)(x - 2)(x + 2) = 0$
$\Rightarrow x = 2, 2, -2$

$\Rightarrow$ track and trail meet at $x = -2 \Rightarrow y = 5(-2) - 9 = -19$ hence $(-2, -19)$

b) (As in item **10·7**, the limits of integration are the x-coordinates of the points of intersection of the two curves involved, and we must integrate (the upper curve − the lower curve))

Question and **Answer** continued ➢

Question and Answer continued

$$\text{Area} = \int_{-2}^{2} [(x^3 - 2x^2 + x - 1) - (5x - 9)]dx = \int_{-2}^{2} (x^3 - 2x^2 - 4x + 8)dx$$

It is better to simplify the integrand and integrate 4 terms rather than 6, but take care with the signs. Most mistakes here are made with simple algebra.

$$\text{Area} = \left[\frac{x^4}{4} - 2\left(\frac{x^3}{3}\right) - 2x^2 + 8x\right]_{-2}^{2} = \left[4 - \frac{16}{3} - 8 + 16\right] - \left[4 + \frac{16}{3} - 8 - 16\right] = 21\frac{1}{3}\text{ units}^2$$

18·5 (Compound and Multiple Angles)

Question and Answer

AD is a diameter of a circle.
B and C lie on the circumference as shown such that
AB = 15 cm, BD = 20 cm, AC = 24 cm.
BD and AC meet at E.
The angles of $\triangle$AED are x, y and z, as shown.

a) Express z in terms of x and y.

b) Calculate the exact value of $\sin z$.

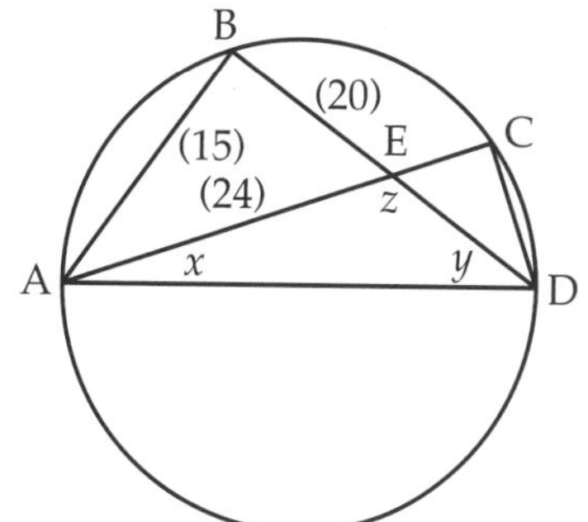

Answer

a) (This is just a hint to get you pointed in the correct direction for part **b)**).
Using the fact that the angles of a triangle add up to 180°, $x + y + z = 180$.
$\Rightarrow z = 180 - (x + y)$.

b) (Even if you cannot see your way through to the end of this question, take the hint from part **a)** and see what use can be made of it.)

$\sin z = \sin[180 - (x + y)]$ (using the result in part **a)**)

$= \sin(x + y)$ (using item **7·3**)

$= \sin x \cos y + \cos x \sin y$ (using item **11·1**)

[or $\sin(x)\cos(y) + \cos(x)\sin(y)$ if you prefer]

It would be nice to have the angles x and y appearing in right-angled triangles, so that we could find their sines and cosines. This is exactly what we do have.

B and C are both angles in a semi-circle, so they are both right angles.
The length of AD is 25 cm (because $\triangle$ABD is a '15, 20, 25'$\triangle$, i.e. 5 times a '3, 4, 5'$\triangle$)
Thus CD = 7 cm (because $\triangle$ACD must be a '7, 24, 25'$\triangle$, (or use Pythagoras' Theorem))

***Question** and **Answer** continued* ➢

Question and Answer continued

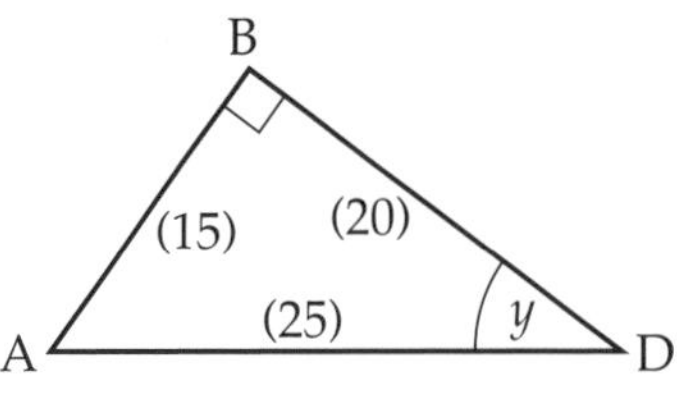

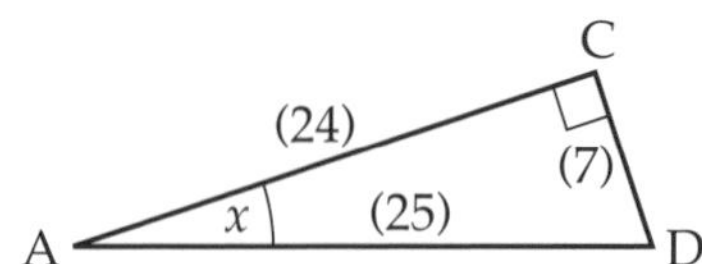

so we have: $\sin x = \frac{7}{25}$, $\cos x = \frac{24}{25}$, $\sin y = \frac{15}{25} = \frac{3}{5}$, $\cos y = \frac{20}{25} = \frac{4}{5}$

$$\therefore \ \sin z = \sin x \cos y + \cos x \sin y$$
$$= \frac{7}{25} \times \frac{4}{5} + \frac{24}{25} \times \frac{3}{5}$$
$$= \frac{28 + 72}{125} = \frac{100}{125} = \frac{4}{5}$$

18·6 (The Circle)

Question and Answer

a) Show that the two circles with equations $x^2 + y^2 + 4x + 6y - 12 = 0$, and $x^2 + y^2 - 20x - 12y + 36 = 0$ touch each other externally, stating the co-ordinates of their point of contact.

b) Find the equation of the smallest circle which can enclose both of these circles.

Answer

(This question contains a bit more than item **12·6**, because I have used the general equation of the circle (so there is more working required to find the centres of the circles and their radii) and I have asked for the common point on the two circles.)

$x^2 + y^2 + 4x + 6y - 12 = 0$ has centre $C(-2, -3)$
and radius $\sqrt{2^2 + 3^2 - (-12)} = \sqrt{25} = 5$

$x^2 + y^2 - 20x - 12y + 36 = 0$ has centre $D(10, 6)$
and radius $\sqrt{(-10)^2 + (-6)^2 - 36} = \sqrt{100} = 10$

so the sum of the radii is $5 + 10 = 15$.

The distance between the centres is CD and

$CD = \sqrt{(10 - (-2))^2 + (6 - (-3))^2} = \sqrt{144 + 81} = \sqrt{225} = 15$

Since the distance between the centres equals the sum of the radii, the circles touch.

A rough sketch is essential for the last part.

Question and Answer continued ➢

Question and Answer continued

The common point divides CD in the ratio 5 : 10.
i.e. P divides CD in the ratio 1 : 2.

10 D (10, 6)
P
5
C (−2, −3)

now $\overrightarrow{CD} = \mathbf{d} - \mathbf{c} = \begin{pmatrix}10\\6\end{pmatrix} - \begin{pmatrix}-2\\-3\end{pmatrix} = \begin{pmatrix}12\\9\end{pmatrix}$

and $\overrightarrow{CP} = \frac{1}{3}\overrightarrow{CD} = \frac{1}{3}\begin{pmatrix}12\\9\end{pmatrix} = \begin{pmatrix}4\\3\end{pmatrix}$

so $\mathbf{p} = \overrightarrow{OP} = \overrightarrow{OC} + \overrightarrow{CP} = \begin{pmatrix}-2\\-3\end{pmatrix} + \begin{pmatrix}4\\3\end{pmatrix} = \begin{pmatrix}2\\0\end{pmatrix}$

i.e. the common point is (2, 0)

b) (Another sketch is called for.)
This shows that RS is the diameter of the required circle, where R and S are as shown.
So Q, the centre of the circle is the mid point of RS.
So we need to find the co-ordinates of R and S.
R is $(-6, -6)$ (since C is the mid point of PR).
S is (18, 12) (since D is the mid point of PS).
Hence Q is (6, 3),
and so the radius $= SQ$
$= \sqrt{(18-6)^2 + (12-3)^2}$
$= \sqrt{225}$
Thus, the outer circle has equation
$(x-6)^2 + (y-3)^2 = 225$
or $x^2 + y^2 - 12x - 6y - 180 = 0$

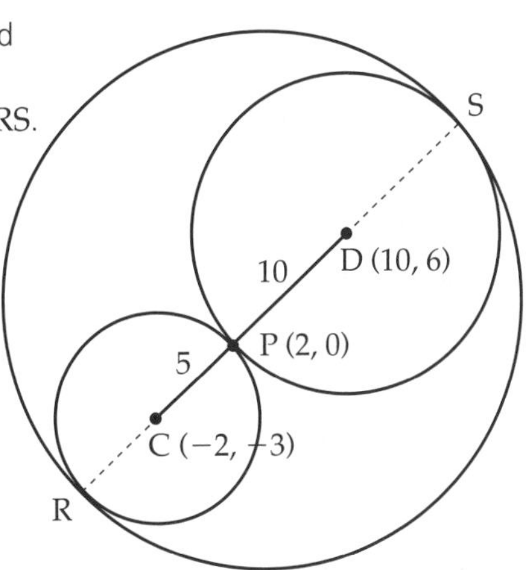

18·7 **(The Circle)**

Question and Answer

A parabola has equation $y = 6 + 4x - x^2$ and a circle has equation $x^2 + y^2 - 4x - 2y - 15 = 0$.

a) Find the equation of the tangent to the parabola at the point where $x = 1$ and show that it is also a tangent to the circle, stating the co-ordinates of the point of contact.

b) Find the equation of the tangent to the circle at the point (6, 3) and show that it is also a tangent to the parabola, stating the co-ordinates of the point of contact.

Answer

a) (We can obtain the equation of the tangent as in item **5·6**. We can then use our knowledge of quadratic theory as in item **8·7** to prove that the line is a tangent to the circle.)

Question and **Answer** continued ➤

Question and Answer continued

$x = 1 \Rightarrow y = 6 + 4x - x^2 = 6 + 4 - 1 = 9 \Rightarrow (1, 9)$
$y = 6 + 4x - x^2 \Rightarrow y' = 4 - 2x \Rightarrow m_{x=1} = 4 - 2 = 2$
so the tangent is the line passing through $(1, 9)$ with gradient 2,
and thus its equation is $y - 9 = 2(x - 1) \Rightarrow y = 2x + 7$

The point of contact with the circle is found by substituting

$$y = 2x + 7 \text{ into } x^2 + y^2 - 4x - 2y - 15 = 0$$

(giving): $x^2 + (2x + 7)^2 - 4x - 2(2x + 7) - 15 = 0$
$\Rightarrow$ $x^2 + 4x^2 + 28x + 49 - 4x - 4x - 14 - 15 = 0$
$\Rightarrow$ $5x^2 + 20x + 20 = 0$
$\Rightarrow$ $x^2 + 4x + 4 = 0$
$\Rightarrow$ $(x + 2)^2 = 0$
$\Rightarrow$ $x = -2, -2$

the equal roots $\Rightarrow$ the line is a tangent to the circle
$x = -2 \Rightarrow y = 2(-2) + 7 = 3 \Rightarrow (-2, 3)$ is the point of contact

b) (Notice the difference. We do not use calculus to find this tangent. We can obtain the equation of the tangent as in item **12·5**.)

The circle has centre (2, 1).
The point (6, 3) is on the circumference.

(6, 3)
(2, 1)

The gradient of the radius $= \dfrac{3 - 1}{6 - 2} = \dfrac{2}{4} = \dfrac{1}{2}$
$\Rightarrow$ the gradient of the tangent is -2
$\Rightarrow$ the equation of the tangent is $y - 3 = -2(x - 6)$
which simplifies to $y = 15 - 2x$

substituting $y = 15 - 2x$ into $y = 6 + 4x - x^2$

gives $15 - 2x = 6 + 4x - x^2$
$\Rightarrow$ $x^2 - 6x + 9 = 0$
$\Rightarrow$ $(x - 3)^2 = 0$
$\Rightarrow$ $x = 3, 3$ and equal roots $\Rightarrow$ the line is a tangent to the parabola
put $x = 3$ in $y = 15 - 2x \Rightarrow y = 15 - 2(3) = 9 \Rightarrow (3, 9)$ is the point of contact

18·8 (Vectors, etc.)

Question and Answer

A room in an art gallery measures $20\,\text{m} \times 30\,\text{m}$ with a height of $4\,\text{m}$.

There is a window built centrally into the roof in the shape of an isosceles triangular prism, measuring $10\,\text{m} \times 20\,\text{m}$ on the base with a height of $2\,\text{m}$.

Security light beams are directed from the end of the window range into the two farthest away corners.

The security light beams are directed from C towards A and B.

Co-ordinate axes are taken as shown.

a) Write down the co-ordinates of A, B and C.

b) Calculate the size of the angle between these two security light beams.

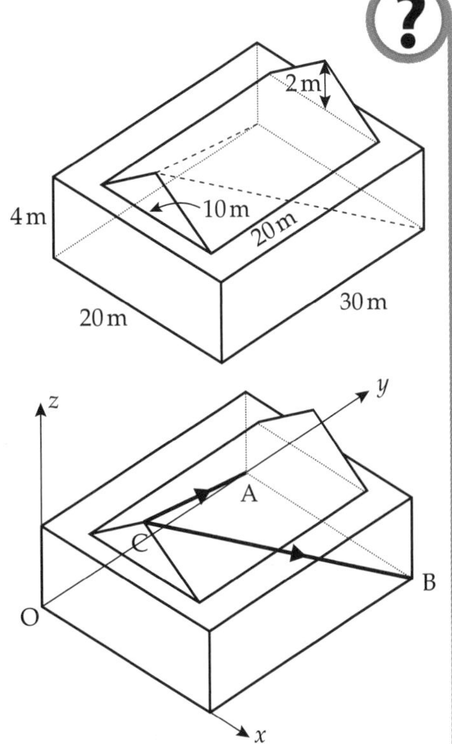

Answer

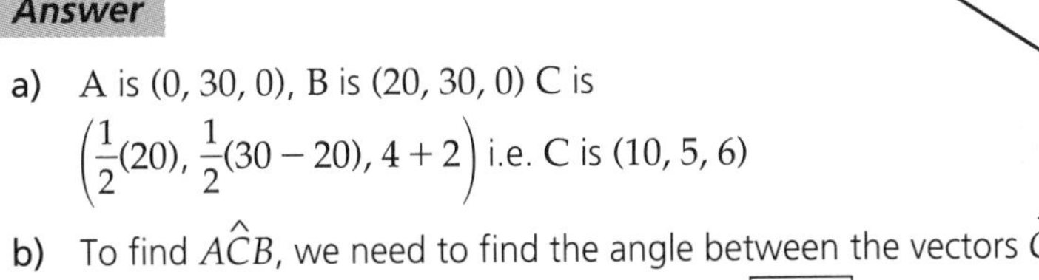

a) A is (0, 30, 0), B is (20, 30, 0) C is $\left(\frac{1}{2}(20), \frac{1}{2}(30-20), 4+2\right)$ i.e. C is (10, 5, 6)

b) To find $A\hat{C}B$, we need to find the angle between the vectors $\overrightarrow{CA}$ and $\overrightarrow{CB}$.

The question then becomes similar to item **13·12** but we must state an appropriate formula relevant to this question, not just a generalisation, i.e.

$$\overrightarrow{CA}.\overrightarrow{CB} = |\overrightarrow{CA}| \times |\overrightarrow{CB}| \times \cos(A\hat{C}B) \Rightarrow \cos(A\hat{C}B) = \frac{\overrightarrow{CA}.\overrightarrow{CB}}{|\overrightarrow{CA}| \times |\overrightarrow{CB}|}$$

We need to make three calculations before we can continue:

$$\overrightarrow{CA} = \mathbf{a} - \mathbf{c} = \begin{pmatrix} 0 \\ 30 \\ 0 \end{pmatrix} - \begin{pmatrix} 10 \\ 5 \\ 6 \end{pmatrix} = \begin{pmatrix} -10 \\ 25 \\ -6 \end{pmatrix} \Rightarrow \text{CA} = \sqrt{100+625+36} = \sqrt{761}$$

$$\overrightarrow{CB} = \mathbf{b} - \mathbf{c} = \begin{pmatrix} 20 \\ 30 \\ 0 \end{pmatrix} - \begin{pmatrix} 10 \\ 5 \\ 6 \end{pmatrix} = \begin{pmatrix} 10 \\ 25 \\ -6 \end{pmatrix} \Rightarrow \text{CB} = \sqrt{100+625+36} = \sqrt{761}$$

$$\overrightarrow{CA}.\overrightarrow{CB} = \begin{pmatrix} -10 \\ 25 \\ -6 \end{pmatrix} . \begin{pmatrix} 10 \\ 25 \\ -6 \end{pmatrix} = -100 + 625 + 36 = 561$$

$$\Rightarrow \cos(A\hat{C}B) = \frac{\overrightarrow{CA}.\overrightarrow{CB}}{|\overrightarrow{CA}| \times |\overrightarrow{CB}|} = \frac{561}{\sqrt{761}\sqrt{761}} = \frac{561}{761} \Rightarrow A\hat{C}B = 42{\cdot}5°$$

18·9 (Logarithms)

Question and Answer

A cup of tea cools according to the law $T_t = T_0e^{-kt}$, where T_0(°C) is the initial temperature and T_t(°C) is the temperature after t minutes.

a) If a cup of tea cools from boiling (100 °C) to 84 °C in 2 minutes, calculate the value of k.

b) To what temperature will the tea have dropped after 5 minutes?

c) I consider tea to be undrinkable below 50 °C. How long does tea take to cool from boiling to my undrinkable temperature?

Answer

a) Substitute values: $84 = 100 \times e^{-k\times 2}$
isolate the exponential term: $0{\cdot}84 = e^{-2k}$
change index form to log form: $\ln 0{\cdot}84 = -2k$
make k the subject: $k = -\frac{1}{2}\log_e 0{\cdot}84 \Rightarrow k = 0{\cdot}0872\ldots$*
*keep this number (unrounded) in one of the memories in your calculator for **b)**

b) (Logarithms are not required for this part. We can either calculate 5 minutes after a temperature of 100 °C or 3 minutes after a temperature of 84 °C, i.e.)

Either:
$T_5 = 100e^{-0{\cdot}0872\times 5} = 64{\cdot}7\,°\text{C}$

or:
$T_3 = 84e^{-0{\cdot}0872\times 3} = 64{\cdot}7\,°\text{C}$

c) We have a similar choice here of starting at 100 °C or 84 °C, but if we choose the latter we must remember to add on the 2 minutes from part **a)**. The working is almost identical to that in part **a)**.

	Either:	or:
Substitute the values:	$50 = 100 \times e^{-0{\cdot}0872t}$	$50 = 84 \times e^{-0{\cdot}0872t}$
isolate the exponential term:	$0{\cdot}5 = e^{-0{\cdot}0872t}$	$0{\cdot}595 = e^{-0{\cdot}0872t}$
change index form to log form:	$\ln 0{\cdot}5 = -0{\cdot}0872t$	$\ln 0{\cdot}595 = -0{\cdot}0872t$
make t the subject:	$t = \frac{\log_e 0{\cdot}5}{-0{\cdot}0872}$	$t = \frac{\log_e 0{\cdot}595}{-0{\cdot}0872}$
	$\Rightarrow t = 7{\cdot}95$ minutes	$\Rightarrow t = 5{\cdot}95$ minutes
		$\Rightarrow$ total time $= 7{\cdot}95$

i.e. 7 minutes 57 seconds

18·10 (Auxiliary Angle)

Question and Answer

A factory fire alarm has a frequency f cycles per second, which varies according to the formula $f = 700 + 100\sin(180t)° + 100\sqrt{3}\cos(180t)°$ where t seconds is the length of time for which it has been switched on.
It has been found that when the frequency drops below 600 c/s, the sound fails to penetrate the ear protectors worn by the machine operatives.
For what percentage of the time is the fire alarm inaudible to these workers.

Answer

(This would be regarded as a difficult question, principally because you are given no guidance with where to start and also because of the 180 in front of t.
You need to spot that f is of the form $700 + a\sin x + b\cos x$.
This suggests writing it in the form $700 + k\sin(x + \alpha)$ [or $700 + k\cos(x - \alpha)$]
I will do this my short way, since the exact values are useful here, but you may prefer to set out your working as in item **16·2** (method 1). As a simplification, I will omit all the degree signs.)

$f = 700 + 100\sin(180t) + 100\sqrt{3}\cos(180t).$

$= 700 + 200\left[\sin(180t) \times \frac{100}{200} + \cos(180t) \times \frac{100\sqrt{3}}{200}\right]$

$= 700 + 200\,[\sin(180t) \times \cos 60 + \cos(180t) \times \sin 60]$

$= 700 + 200\sin(180t + 60)$

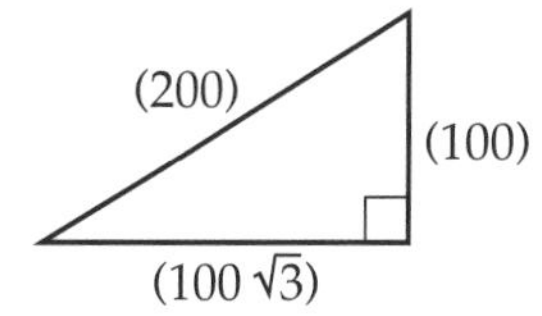

In this form it is much easier to observe the properties of f. A quick sketch of the graph of the function should help us with the remaining part of the problem.

$f_{max} = 700 + 200 \times 1 = 900$: when $180t + 60 = 90$; i.e. $180t = 30 \Rightarrow t = \frac{1}{6}$

$f_{min} = 700 + 200 \times (-1) = 500$: when $180t + 60 = 270$; i.e. $180t = 210 \Rightarrow t = \frac{7}{6}$

$f_{mid} = 700 + 200 \times 0 = 700$: when $180t + 60 = 180, 360$; i.e. $180t = 120, 300 \Rightarrow t = \frac{2}{3}, \frac{5}{3}$

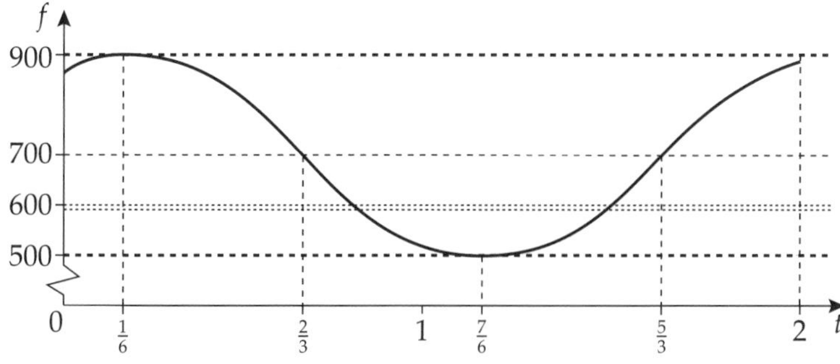

Question and Answer continued ➢

Question and Answer continued

We need to calculate the length of time in each cycle (of 2 seconds) for which the frequency drops below 600 c/s.

$$f = 600 \Rightarrow 700 + 200\sin(180t + 60) = 600$$

$$\Rightarrow \sin(180t + 60) = -\frac{1}{2}$$

$$\Rightarrow 180t + 60 = 210, 330 \Rightarrow t = \frac{5}{6}, \frac{3}{2}$$

$\Rightarrow$ from $t = \frac{5}{6}$ till $t = \frac{3}{2}$ is $\frac{3}{2} - \frac{5}{6} = \frac{2}{3}$ seconds

i.e. below 600 c/s for $\frac{2}{3}$ s out of every 2 s i.e. $33\frac{1}{3}\%$

18·11 (Differentiation, Functions and Graphs)

Question and Answer

a) Sketch the curve with equation $y = f(x) = 2x^3 + 3x^2$.

b) Hence sketch the graph of $y = g(x) = 2 - f(x)$.

c) Find where $y = g(x)$ crosses the x-axis correct to two decimal places.

d) Find the size of the angle which the tangent to $y = g(x)$ at the point where $x = \frac{1}{2}$ makes with the x-axis.

Answer

a) (Look for intersections with both axes, stationary points, symmetry, behaviour for large values of x.)

$x = 0 \Rightarrow y = 2(0)^3 + 3(0)^2 = 0$

$y = 0 \Rightarrow 2x^3 + 3x^2 = 0 \Rightarrow x^2(2x + 3) = 0 \Rightarrow x = 0, 0, -\frac{3}{2}$

so the only intersections with the axes are at (0, 0), $\left(-\frac{3}{2}, 0\right)$.

for stationary points, $y' = 0$

$\Rightarrow 6x^2 + 6x = 0 \Rightarrow 6x(x + 1) = 0 \Rightarrow x = -1, 0$

$\Rightarrow y = 1, 0$

x	-1^-	-1	-1^+	0^-	0	0^+
$6x$	–	–	–	–	0	+
$(x+1)$	–	0	+	+	+	+
$6x(x+1)(= y')$	+	0	–	–	0	+
(tangent)	↗	→	↘	↘	→	↗

alternatively:

$y'' = 12x + 6$

$\Rightarrow y''(-1) = -6 < 0$

$\Rightarrow (-1, 1)$ is a max. t. pt.

and $y''(0) = 6 > 0$

$\Rightarrow (0, 0)$ is a min. t. pt.

so $(-1, 1)$ is a max. t. pt., (0, 0) is a min. t. pt.

Question and Answer continued ➤

Question and Answer continued

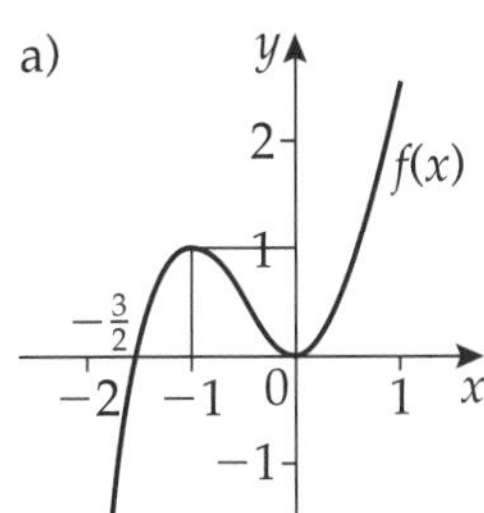

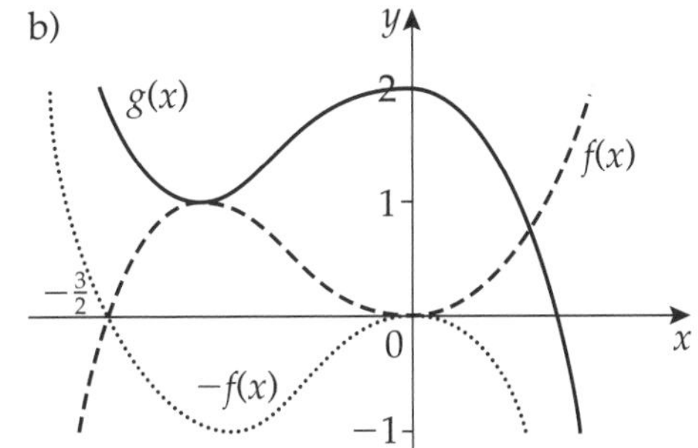

b) Think of $g(x)$ as $-f(x)+2$, i.e. reflect $f(x)$ in the x-axis then translate 2 up.

c) (This is calculator business): $y = g(x) = 2 - f(x) = 2 - (2x^3 + 3x^2) = 2 - 3x^2 - 2x^3$.

$g(0) = 2 (> 0)$	$g(0{\cdot}6) = 0{\cdot}488 (> 0)$	$g(0{\cdot}67) = 0{\cdot}052 (> 0)$
$g(1) = -3 (< 0)$	$g(0{\cdot}7) = -0{\cdot}156 (< 0)$	$g(0{\cdot}68) = -0{\cdot}016 (< 0)$

Now try half way between 0·67 and 0·68 to decide which.

$g(0{\cdot}675) = 0{\cdot}018 > 0$

Compare with item **9·8**

hence $x = 0{\cdot}68$

d) $g(x) = 2 - 3x^2 - 2x^3 \Rightarrow g'(x) = -6x - 6x^2 \Rightarrow g'\left(\frac{1}{2}\right) = -3 - 1{\cdot}5 = -4{\cdot}5$

$\Rightarrow$ gradient of tangent $= -4.5$

Compare item **3·3** – tan (angle) $\Rightarrow$ angle $= 180° - \tan^{-1}(4{\cdot}5) = 102{\cdot}5°$

18·12 (Vectors)

Question and Answer

ABCD, EFGH is a cuboid.
K is the mid point of HG.
L divides BC in the ratio 1 : 2.
Relative to suitable axes,
$\overrightarrow{AB} = 4\mathbf{i} + 2\mathbf{j} + 2\mathbf{k}$
$\overrightarrow{AD} = -3\mathbf{i} + 6\mathbf{j} + 3\mathbf{k}$, and
$\overrightarrow{AE} = -6\mathbf{i} - 18\mathbf{j} + 30\mathbf{k}$.

Find the components of $\overrightarrow{AK}$, $\overrightarrow{AL}$ and $\overrightarrow{KL}$.

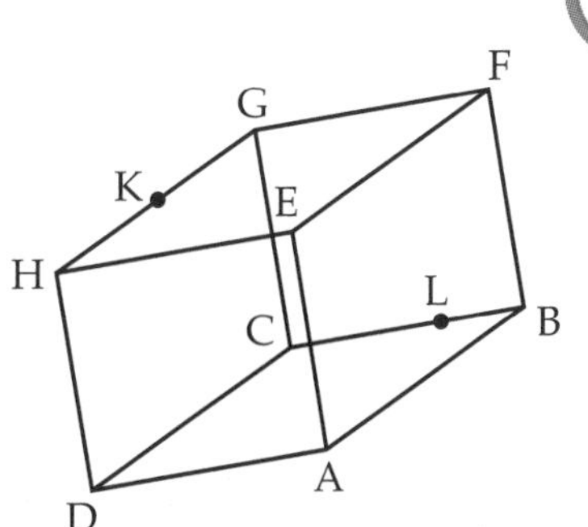

Question and Answer continued ➢

Question and Answer continued

Answer

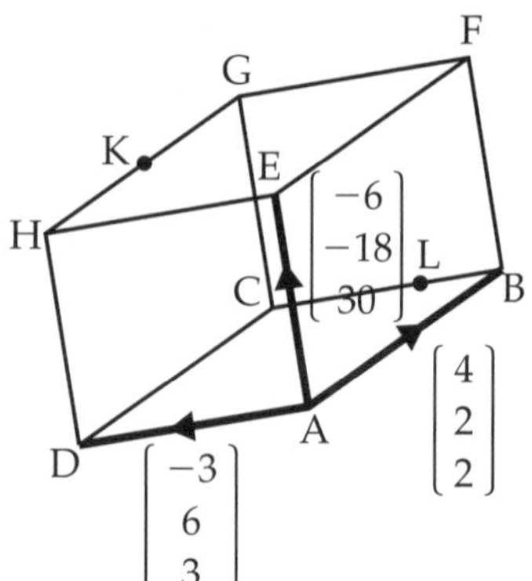

Let us use column vectors, and label our diagram.
We can find pathways from A to K and L.

$$\overrightarrow{AK} = \overrightarrow{AD} + \overrightarrow{DH} + \overrightarrow{HK}$$

$$= \overrightarrow{AD} + \overrightarrow{AE} + \frac{1}{2}\overrightarrow{HG}$$

$$= \overrightarrow{AD} + \overrightarrow{AE} + \frac{1}{2}\overrightarrow{AB}$$

$$= \begin{pmatrix}-3\\6\\3\end{pmatrix} + \begin{pmatrix}-6\\-18\\30\end{pmatrix} + \begin{pmatrix}2\\1\\1\end{pmatrix} = \begin{pmatrix}-7\\-11\\34\end{pmatrix}$$

$$\overrightarrow{AL} = \overrightarrow{AB} + \overrightarrow{BL}$$

$$= \overrightarrow{AB} + \frac{1}{3}\overrightarrow{BC}$$

$$= \overrightarrow{AB} + \frac{1}{3}\overrightarrow{AD}$$

$$= \begin{pmatrix}4\\2\\2\end{pmatrix} + \begin{pmatrix}-1\\2\\1\end{pmatrix} = \begin{pmatrix}3\\4\\3\end{pmatrix}$$

$$\overrightarrow{KL} = \overrightarrow{KA} + \overrightarrow{AL}$$

$$= \overrightarrow{AL} - \overrightarrow{AK}$$

$$= \begin{pmatrix}3\\4\\3\end{pmatrix} - \begin{pmatrix}-7\\-11\\34\end{pmatrix} = \begin{pmatrix}10\\15\\-31\end{pmatrix} \text{ (or } \overrightarrow{KL} = 10\mathbf{i} + 15\mathbf{j} - 31\mathbf{k}\text{)}$$

18·13 (Further Calculus)

Question and Answer

a) Differentiate $\cos^3 x$ with respect to x.

b) Hence evaluate $\int_0^{\pi} (\sin x \cos^2 x)dx$.

Answer

a) Think of $\cos^3 x$ as $(\cos x)^3$; then to differentiate a 'bracket cubed', you get three times the 'bracket squared' times the derivative of the bracket.

$$\frac{d}{dx}(\cos^3 x) = \frac{d}{dx}(\cos x)^3 = 3(\cos x)^2(-\sin x) = -3\sin x \cos^2 x$$

Compare this with item 14·2b.

b) (You can only do part **b)** if you remember that integration and differentiation are inverse processes and if you notice that what you are being asked to integrate is what you obtained for your answer to the differentiation in part **a)** divided by (−3). You have no other way of being able to complete this question.)

$$\int_0^{\pi} (\sin x \cos^2 x)dx = -\frac{1}{3}\int_0^{\pi} (-3\sin x \cos^2 x)dx$$

$$= -\frac{1}{3}\Big[\cos^3 x\Big]_0^{\pi} \qquad \text{(using part a)}$$

$$= -\frac{1}{3}\Big[(-1)^3 - (1)^3\Big] = \frac{2}{3} \qquad \text{(note that when } x = 0\text{, the bracket} \neq 0)$$

18·14 (Compound and Multiple Angles, Integration)

Question and Answer

a) Express $\cos^2 2x$ in terms of $\cos 4x$.

b) Hence evaluate $\int_0^{\frac{\pi}{4}} (\cos^2 2x)dx$.

Answer

a) (You can write the answer down in one line if you can remember the last two formulae in item 11·1. Failing that you can work out the answer from one of the formulae for $\cos(2A)$; select the one which involves $\cos^2 A$.

$$\cos(2A) = 2\cos^2(A) - 1 \quad \text{let } A = 2x \Rightarrow \cos(4x) = 2\cos^2(2x) - 1$$

$$\Rightarrow \cos(4x) + 1 = 2\cos^2(2x)$$

$$\Rightarrow \cos^2(2x) = \frac{1}{2}(\cos(4x) + 1)$$

Question and **Answer** continued ➢

Question and Answer continued

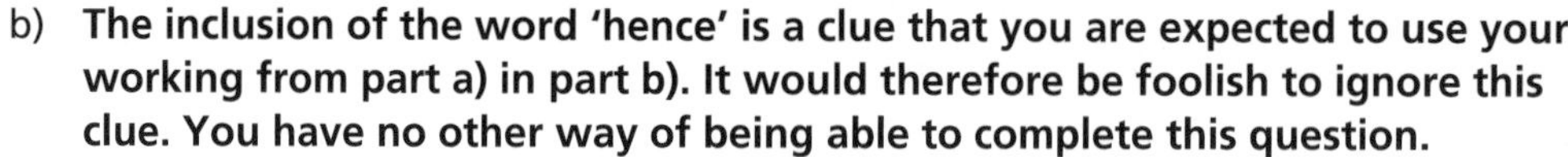

b) **The inclusion of the word 'hence' is a clue that you are expected to use your working from part a) in part b). It would therefore be foolish to ignore this clue. You have no other way of being able to complete this question.**

$$\int_0^{\frac{\pi}{4}}(\cos^2 2x)dx = \frac{1}{2}\int_0^{\frac{\pi}{4}}(1+\cos 4x)dx = \frac{1}{2}\left[x+\frac{\sin 4x}{4}\right]_0^{\frac{\pi}{4}} = \frac{1}{2}\left\{\left[\frac{\pi}{4}+0\right]-[0+0]\right\} = \frac{\pi}{8}$$

18·15 **(Differentiation)**

Question and Answer

Find the stationary points on the curve with equation $y = x^3(x^2 - 15)$

Answer

$y = x^3(x^2 - 15) = x^5 - 15x^3$

$\Rightarrow \dfrac{dy}{dx} = 5x^4 - 45x^2 = 5x^2(x^2 - 9) = 5x^2(x-3)(x+3)$;

for stationary points $\dfrac{dy}{dx} = 0$, so $x = -3, 0, 0, 3$

so $y = 162, 0, 0, -162$

x	-3^-	-3	-3^+	0^-	0	0^+	3^-	3	3^+
$(x+3)$	$-$	0	+	+	+	+	+	+	+
$5x^2$	+	+	+	+	0	+	+	+	+
$(x-3)$	$-$	$-$	$-$	$-$	$-$	$-$	$-$	0	+
$\frac{dy}{dx} = 5x^2(x-3)(x+3)$	+	0	$-$	$-$	0	$-$	$-$	0	+
(tangent)	↗	→	↘	↘	→	↘	↘	→	↗

Hence, $(-3, 162)$ max. t. pt., $(0, 0)$ inflection, $(3, -162)$ min. t. pt.

(The statement that 'As $5x^2$ is always positive, the sign of y' depends only on the sign of $(x^2 - 9)$.' would be an acceptable (but wordier) replacement for the third row in this nature table.)

(Alternatively):

$y''(x) = 20x^3 - 90x = 10x(2x^2 - 9)$

$\Rightarrow y''(-3) = -270(<0) \Rightarrow (-3, 162)$ is a max.t. pt.

$y''(3) = 270(>0) \Rightarrow (3, 162)$ is a max.t. pt.

$y''(0) = 0 \Rightarrow$ we still need a nature table for this bit!

18·16 (Differentiation)

Question and Answer

a) Sketch the curve with equation $y = x^3(x^2 - 15)$.
b) For what values of k does the equation $x^3(x^2 - 15) = k$ have three real roots?

Answer

a) This is the same curve as **18·15**, so we already know the turning points and their nature: $(-3, 162)$ max. t. pt. $(0, 0)$ inflection $(3, -162)$ min. t. pt.

intersections with axes:

$x = 0 \Rightarrow y = 0$
$y = 0 \Rightarrow x^3(x^2 - 15) = 0 \Rightarrow x = 0, 0, 0, \pm\sqrt{15}$
hence $(0, 0)$, $(-\sqrt{15}, 0)$, $(\sqrt{15}, 0)$

symmetry:
This is an odd function.
So it has half turn symmetry about the origin.

behaviour for large x:
x^5 is the leading term
$x \to -\infty \Rightarrow y \to (-\infty)^5 \to -\infty$
$x \to +\infty \Rightarrow y \to (+\infty)^5 \to +\infty$

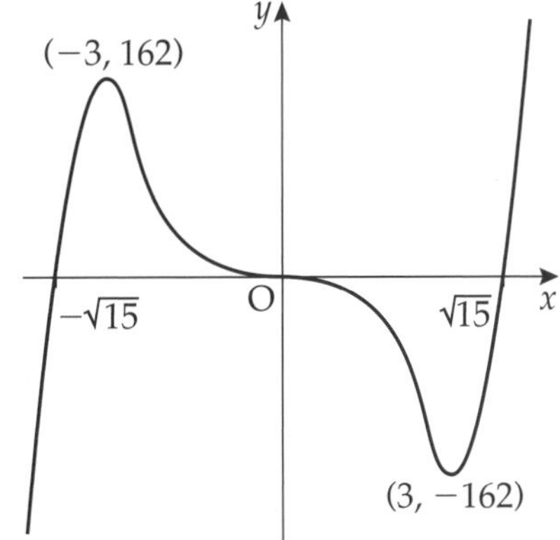

b) The solutions of the equation $x^3(x^2 - 15) = k$ must be the simultaneous solutions of $y = x^3(x^2 - 15)$ and $y = k$
On this graph, think of moving the straight line up and down (changing the value of k) and see how often it intersects the curve. It becomes clear that the answer is

$-162 < k < 162$.

[Additionally, it would have two distinct real roots for $k = \pm162$.]

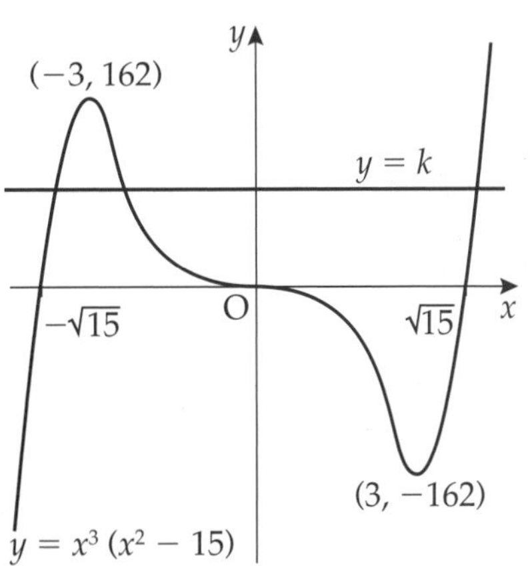

18·17 (The Straight Line, The Circle)

Question and Answer

Find the equation of the circle which passes through the points P(−3, 8), Q(3, 10) and R(9, 2).

Answer

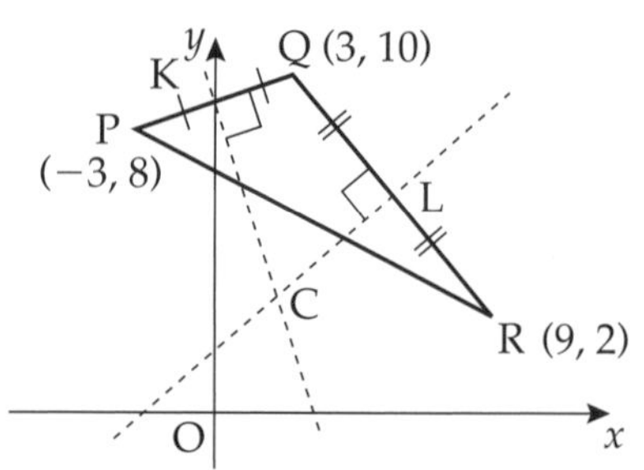

The centre, C, of circle PQR is the point of concurrence of the perpendicular bisectors of the chords PQ, QR and PR. [See page 15].

We only need to find the equations of two of these lines and hence where they intersect. None seems any simpler than the others, so at random, I will choose the perpendicular bisectors of PQ and QR.

Let K and L be the mid points of PQ and QR.

Introducing your own letters like this simplifies your writing out of the solution and makes it more intelligible to the marker. It also helps you if you need to check through for an error.

to find the equation of KC:

K is (0, 9), $m_{PQ} = \frac{10-8}{3-(-3)} = \frac{2}{6} = \frac{1}{3} \Rightarrow m_{KC} = -3$

$\Rightarrow$ KC has equation $y - 9 = -3(x - 0) \Rightarrow y = 9 - 3x$

to find the equation of LC:

L is (6, 6), $m_{QR} = \frac{10-2}{3-9} = \frac{8}{-6} = -\frac{4}{3} \Rightarrow m_{LC} = \frac{3}{4}$

$\Rightarrow$ LC has equation $y - 6 = \frac{3}{4}(x - 6) \Rightarrow 4y - 24 = 3x - 18$

$\Rightarrow 4y = 3x + 6$

solving KC and LC simultaneously:

substituting $9 - 3x$ for y in LC gives $4(9 - 3x) = 3x + 6$

$\Rightarrow 36 - 6 = 3x + 12x$

$30 = 15x \Rightarrow x = 2$

substituting $x = 2$ in KC gives $y = 9 - 3x = 9 - 6 = 3$ hence C is (2, 3)

to calculate the length of the radius, r:

$r = \text{CP} \Rightarrow r^2 = (-3-2)^2 + (8-3)^2 = 25 + 25 = 50$

[It is worthwhile checking that $CQ^2 = CR^2 = 50$ as well.
If this is the case, it is unlikely that you will have made an error]

now using $(x - a)^2 + (y - b)^2 = r^2$

the equation of the circle is $(x - 2)^2 + (y - 3)^2 = 50$ or $x^2 + y^2 - 4x - 6y - 37 = 0$

Now that you have made it to the end of this book, I trust that I have given you some information and insight which will be worth more marks to you in your Higher Mathematics examination. Do your best. Best wishes.

Chapter 19

MULTIPLE CHOICE QUESTIONS

In 2008, the format of the Higher Mathematics examination changed.
The exam consists of two question papers as before:

Paper 1 (Non Calculator)	1 hour 30 minutes	70 Marks
Paper 2 (Calculators Allowed)	1 hour 10 minutes	60 Marks

The main change was in Paper 1. It now consists of two sections, A and B. The changes are summarised below:

Section	**Number of Questions**	**Format**	**Total Marks**
A	20	Multiple Choice	40
B	Approx. 3 to 5	Short and Extended Response	30

The main reason for these changes is to enable the SQA to assess a wider proportion of the syllabus.

Notes on Multiple Choice Questions

- Each multiple choice question (sometimes referred to as Objective Questions) has four options A, B, C and D to choose from.
- For each question there is only **one** correct answer.
- Each question is answered onto a special answer grid provided.
- Your rough working for each question is done on a separate piece of paper – only the final answer on your grid will be marked.
- Each question is worth 2 marks for a correct response – there is no deduction of marks for incorrect answers.
- Although each question has only one correct answer, the remaining three options will not be random answers – they will be based on the most probable mistakes candidates are likely to make (they are sometimes called *distracters*).

Tips for tackling Multiple Choice Questions

- As for any maths question, make sure that you set out your working carefully and check each stage as you go along. You may not receive any credit for this (unlike the rest of the exam) but it is still extremely good practice.
- It's a good idea to avoid looking at the four options before attempting the question – as mentioned before, the three incorrect answers could distract you.
- Make sure that you record only one answer for each question – any questions with two or more options recorded will automatically be marked incorrect.
- Make sure that you answer each question even if you have to guess (as a last resort) – at least you have a one in four chance of guessing the correct answer to that particular question. (Incidentally, the odds of guessing the correct answers to all 20 questions is 1 099 511 627 776 to 1)!

Here is a selection of some typical multiple choice questions.

1

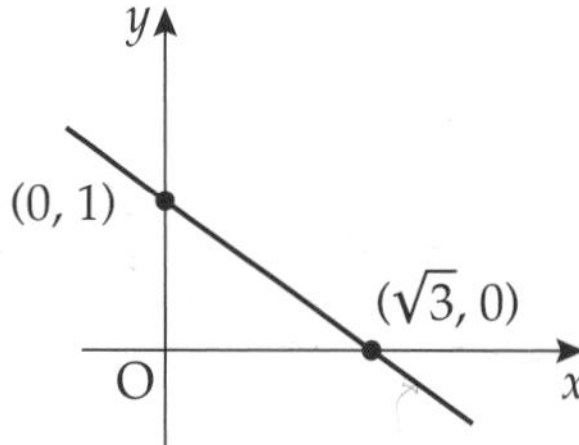

The angle between the line shown and the positive direction of the x-axis is:

A $60°$ B $120°$ C $135°$ D $150°$

2 Triangle ABC has vertices A(3, 5), B(4, −2) and C(−2, 3)

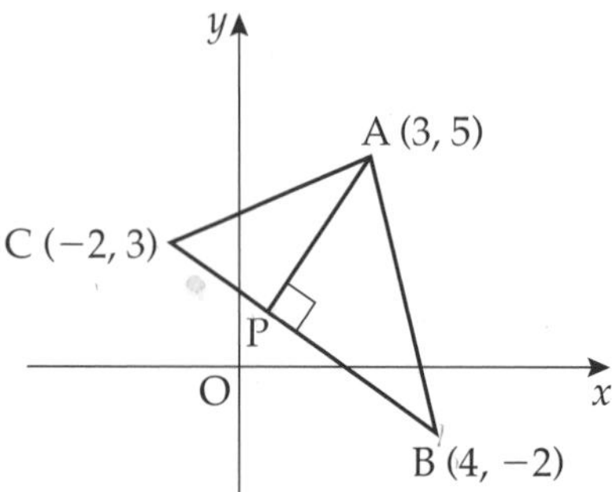

The gradient of altitude AP is:

A $-\frac{5}{6}$ B $\frac{5}{6}$ C $-\frac{6}{5}$ D $\frac{6}{5}$

3 Two functions are defined as:

$$f(x) = x^2 - 1 \text{ and } g(x) = \frac{1}{x} \text{ where } x \neq 0.$$

$f(g(x))$ equals:

A $(x^2 - 1)\frac{1}{x}$ B $\frac{1}{x^2} - 1$ C $\frac{1}{x^2 - 1}$ D $\frac{1}{x^2} - x$

4 The graph of the function $y = f(x)$ is shown.

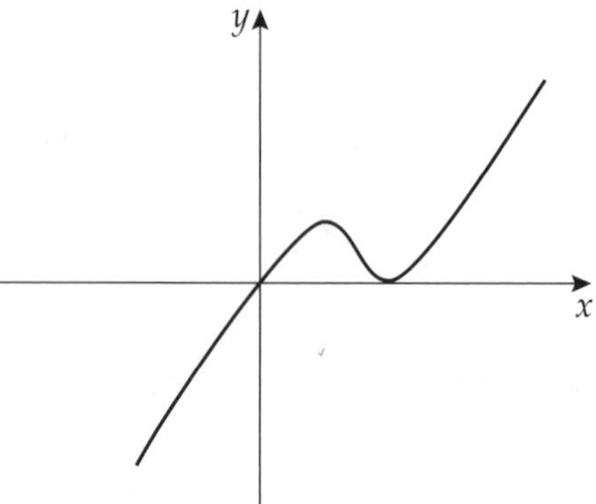

Which of the following graphs could represent $y = 2 - f(x)$?

A

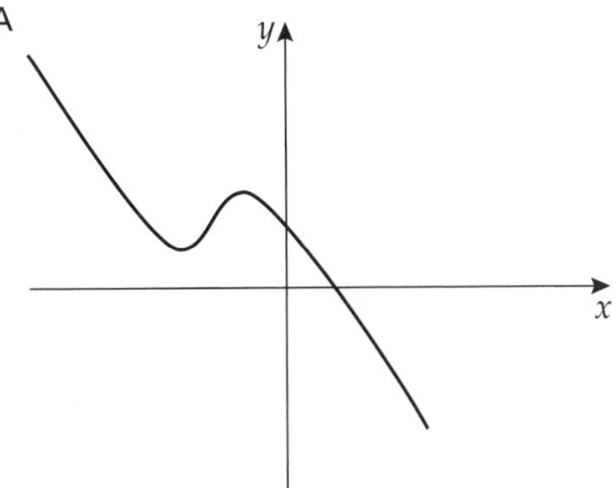

B

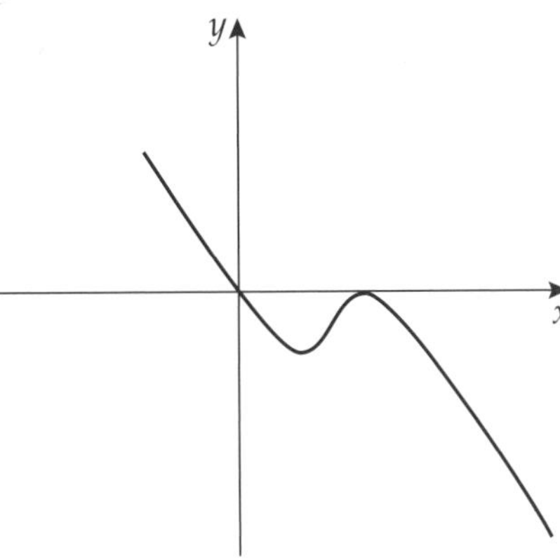

C

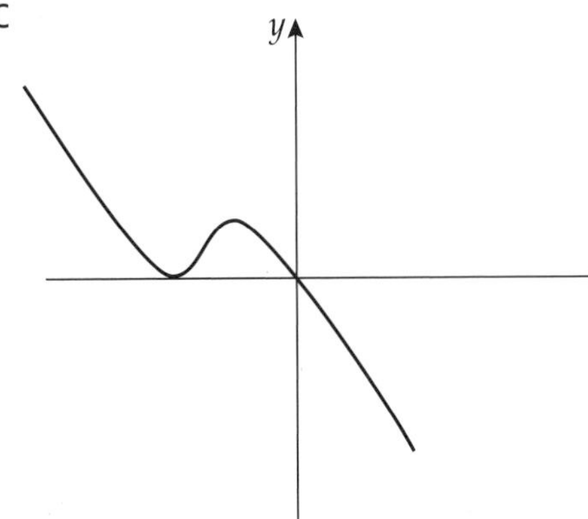

D

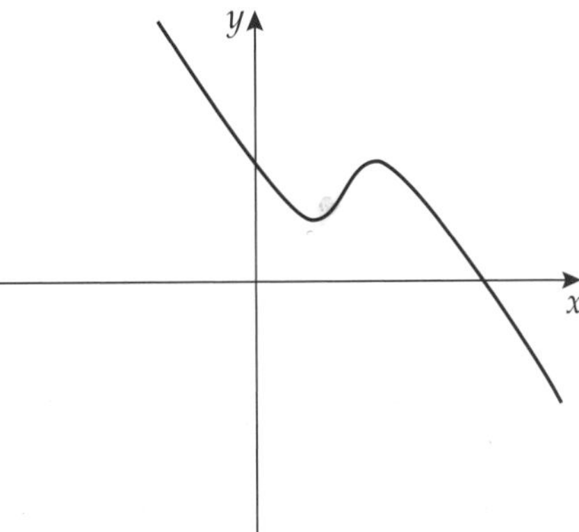

5 A curve has equation $y = 6x - 2x^2$.
The gradient of the tangent to this curve at the point $(1, 4)$ is:

A 4 B -10 C 2 D $\frac{7}{3}$

6 The values of the x-coordinates of the stationary points of the curve $y = 3x^2 - x^3$ are:

A $x = -\sqrt{3}, x = \sqrt{3}$ B $x = 0, x = 2$ C $x = 0, x = -2$ D $x = 2, x = -2$

7 The recurrence relation $u_{n+1} = 2u_n - 3$ has the value $u_0 = -1$.
The value of u_2 is:

A -13 B -5 C -1 D -29

8 The limit of the recurrence relation $u_{n+1} = 0.9u_n - 2$ is:

A -2 B $-\frac{20}{9}$ C 2 D -20

9 The polynomial has $x^3 + 2x^2 - x - 2$ has $(x - 1)$ as one of its factors. The other factors are:

A $(x - 1)$ and $(x + 2)$ B $(x + 1)$ and $(x + 2)$

C $(x - 1)$ and $(x - 2)$ D $(x + 1)$ and $(x - 2)$

10 The equation $2x^2 - x - p = 0$ has equal roots.
The value of p is:

A $\frac{1}{4}$ B $-\frac{1}{4}$ C $\frac{1}{8}$ D $-\frac{1}{8}$

11 Evaluate $\int_1^2 \frac{1}{x^2}\,dx$.

A $\frac{1}{2}$ B $\frac{7}{4}$ C $-\frac{3}{4}$ D $-\frac{1}{2}$

12 $\int \frac{x - \sqrt{x}}{x}\,dx$ equals:

A $1 + 2x^{\frac{1}{2}} + c$ B $x + 2x^{\frac{1}{2}} + c$ C $1 - 2x^{\frac{1}{2}} + c$ D $x - 2x^{\frac{1}{2}} + c$

13 Given that $\cos a° = \frac{2}{3}$ then the exact value of $\cos 2a°$ equals:

A $\frac{4}{3}$ B $-\frac{1}{9}$ C $\frac{1}{3}$ D $-\frac{1}{3}$

14 The exact value of $2\sin\frac{3\pi}{4}$ is:

A $\frac{2}{\sqrt{2}}$ B $-\frac{1}{\sqrt{2}}$ C $\frac{1}{\sqrt{2}}$ D $-\frac{2}{\sqrt{2}}$

15 The circle shown below has its centre at $(6, -1)$ and has the tangent $y = 4$.

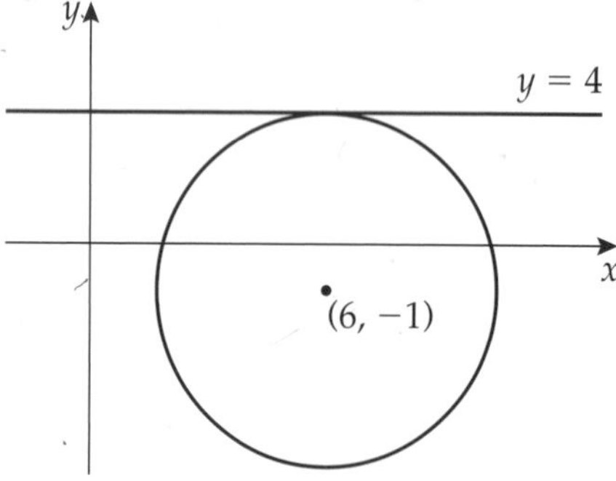

The equation of the circle is:

A $(x - 6)^2 + (y + 1)^2 = 5$ B $(x + 6)^2 + (y - 1)^2 = 25$

C $(x - 6)^2 + (y + 1)^2 = 25$ D $(x + 1)^2 + (y - 6)^2 = 25$

16

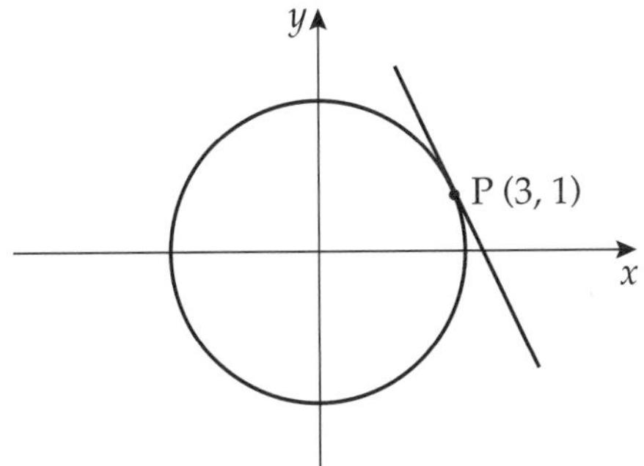

A circle with equation $x^2 + y^2 = 10$ has a tangent at the point P(3, 1). The equation of the tangent at P is:

A $y = -\frac{1}{3}x + 2$ B $y = -3x + 10$ C $y = -3x + 6$ D $y = -3x + 2$

17 The vectors in the diagram below have lengths $|\boldsymbol{a}| = 2$ and $|\boldsymbol{b}| = 3$. The angle between $\boldsymbol{a}$ and $\boldsymbol{b}$ is 30°.

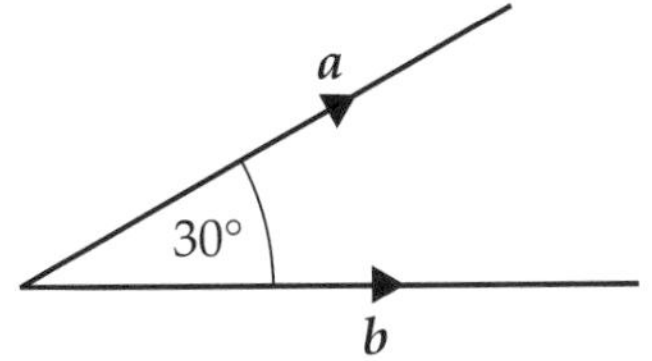

The value of $\boldsymbol{a}.\boldsymbol{b}$ is:

A 6 B 3 C $3\sqrt{3}$ D $\frac{\sqrt{3}}{2}$

18 $UVWX$ is a tetrahedron with $\overrightarrow{UV} = \boldsymbol{a}$, $\overrightarrow{UX} = \boldsymbol{b}$ and $\overrightarrow{XW} = \boldsymbol{c}$.

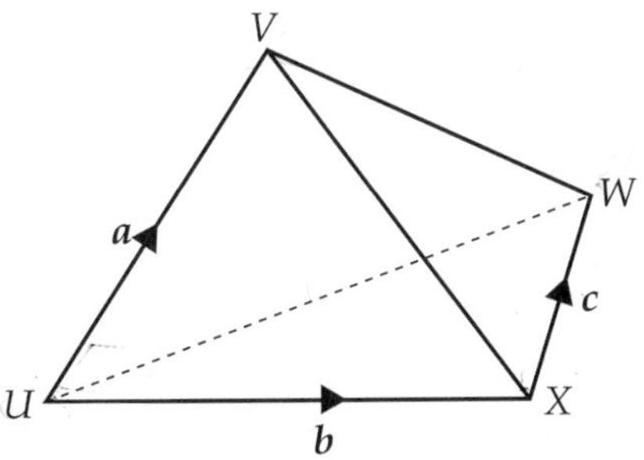

The vector $\overrightarrow{WV}$ equals:

A $\boldsymbol{b} - \boldsymbol{a} + \boldsymbol{c}$ B $-\boldsymbol{b} + \boldsymbol{a} - \boldsymbol{c}$ C $\boldsymbol{c} + \boldsymbol{a} - \boldsymbol{b}$ D $\boldsymbol{c} - \boldsymbol{a} - \boldsymbol{b}$

19 $\int_0^{\frac{\pi}{2}} \cos 3x \, dx$ equals

A -1 B $\frac{1}{3}$ C 1 D $-\frac{1}{3}$

20 If $y = \sqrt{x^2 - 1}$, then $\frac{dy}{dx}$ equals:

A $x(x^2 - 1)^{-\frac{1}{2}}$ B $2x(x^2 - 1)^{-\frac{1}{2}}$ C $\frac{1}{2}(x^2 - 1)^{-\frac{1}{2}}$ D $\frac{2}{3}(x^2 - 1)^{\frac{3}{2}}$

21 The graphs of $y = \log_3 x$ and $y = 2$ are shown below.
They intersect at the point P.

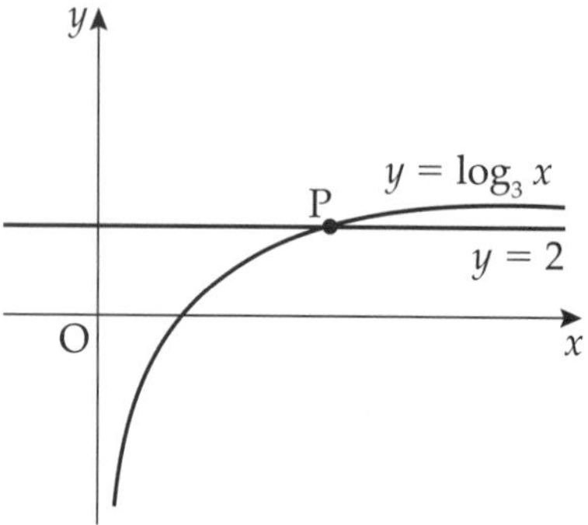

The coordinates of P are:

A $(3, 2)$ B $(9, 2)$ C $(6, 2)$ D $(8, 2)$

22 The solution to the equation $\log_5(x - 1) + 1 = 0$ is:

A $x = \frac{6}{5}$ B $x = 6$ C $x = 26$ D $x = -4$

23 The maximum value of $\cos x + \sqrt{3} \sin x$ is:

A $\sqrt{3}$ B $1 + \sqrt{3}$ C 2 D 1

24 When $3 \sin x° - \sqrt{3} \cos x°$ is expressed in the form $k \sin(x - a)°$, $0° \leqslant a \leqslant 90°$, then the exact value of a is:

A $30°$ B $45°$ C $60°$ D $75°$

25 Below is an isosceles triangle with base parallel to the x-axis.

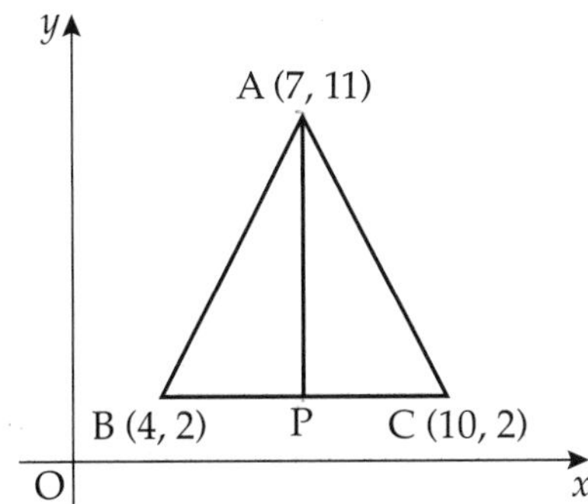

The equation of altitude AP is:

A $x = 11$ B $y = 11$ C $x = 7$ D $y = 7$

26 Which of the following would be suitable domains of the function

$$f(x) = \frac{\sqrt{x}}{x - 1}?$$

A $x \in R, x \neq 1$ B $x \in R, x \neq 0$

C $x \in R, x \neq 1$ and $x \leqslant 0$ D $x \in R, x \neq 1$ and $x \geqslant 0$

27 $2x^2 - 6x - 1$ can be expressed as:

A $2(x - \frac{3}{2})^2 - 1$ B $2(x - \frac{3}{2})^2 - \frac{13}{4}$

C $2(x - 3)^2 - 10$ D $2(x - \frac{3}{2})^2 - \frac{11}{2}$

28 When $2x^3 - 3x^2 - 5$ is divided by $(x + 1)$ the remainder is:

A 0 B -10 C -6 D -7

29 The roots of the quadratic equation $6x^2 + 5x - 4 = 0$ are:

A real, distinct and rational B real and equal

C not real D real and distinct and irrational

30 For the following,

I $x^2 + y^2 + 2x + 4y + 6 = 0$
II $x^2 + y^2 - 2x + 4y + 6 = 0$
III $x^2 + y^2 + 2x - 4y + 6 = 0$

which of these represent the equation of a circle?

A All of **I**, **II** and **III**

B **II** and **III** only

C **I** only

D None of **I**, **II** and **III**

Answers

1	2	3	4	5	6	7	8	9	10	11	12	13	14	15
D	D	B	D	C	B	A	D	B	D	A	D	B	A	C

16	17	18	19	20	21	22	23	24	25	26	27	28	29	30
B	C	B	D	A	B	A	C	A	C	D	D	B	A	D

APPENDIX ONE: EXAMINATION TECHNIQUE

[Remember to read this more than once, but not *all* at once.]

Examination Technique is not a set of tricks to gain you more marks than you deserve in the exam; it is about knowing to do certain things which will give you more chance of scoring marks or avoid losing them. How you work throughout the year has the most bearing on how you perform on the day, but you do not wish to lose any marks through lack of exam sophistication. Most of the advice in this appendix concerns the exam itself and immediately before it, but much of it is also valid for your prelim and any other tests you may be set. The advice is gathered in the appendix under these headings:

Final Preparation for the Exam
Tips for improving your mark
Avoiding Errors
Being helpful to the marker
Some Rules

Final Preparation for the Exam

- Ensure that you have worked through at least three complete recent previous papers (or other practice papers) against the clock, to build your confidence and give you experience of how long you need to tackle the amount of work involved in an exam paper. If there is a type of question or area of the syllabus that you are less good at, select appropriate questions from earlier past papers to help you to improve at this. I am sure that your teacher (or tutor, if you have one) will be delighted that you have identified a weakness and will be only too willing to assist you to rectify this gap in your knowledge or skill. As was stated earlier, the more practice you have had, the better you are likely to perform, so more than three past papers would be even better.
- Do not revise excessively the evening before your exam or stay up too late. Have all of your revision completed before this and spend the evening relaxing, e.g. take a walk in the fresh air. You require to be 'up for it', i.e. mentally attuned to sitting your exam and not worried about extraneous issues.
- There is no extension of time allowed for late arrival, so rise promptly and ensure that you arrive at the exam hall in plenty of time, without having had to rush to get there. Make sure of your transport arrangements well before the exam day. If you have been allocated a seat number, it is best to keep a note of this and not have to be scurrying around looking for it at the last minute.
- Bring adequate equipment with you. You are expected to write in black or dark blue ink. DO NOT write your answers in pencil. Under no circumstances write in red or green ink; that is what the markers and checkers do. Make sure that you have more than one pen, in case it runs out of ink. Bring a ruler and your own calculator, which should preferably be a graphics calculator. Many a pupil has borrowed a calculator for the exam and worked consistently with it in the wrong mode. You cannot expect a marker to second guess where your wrong answers came from, so you would be penalised every time that it made

a difference. Familiarity with your own machine is essential, even just for peace of mind. Put new batteries in your calculator, or at least have replacements in your pocket. You do not want it to fail you in the middle of the exam.

Tips for improving your mark

- Show all your working. The front of the question paper tells you that full credit will only be given where there is adequate working shown. If you do three lines in your head and do not communicate how you arrived at that answer, then you may not get full credit. If you are bright enough to do three processes in your head, you are bright enough to know that you must communicate to the marker exactly how these processes work. If you offer the wrong answer with no working, you will get no marks at all. So be warned. A good phrase used in Portree High School, I believe, to reinforce this lesson is to SHOW WHAT YOU KNOW. Showing all your working is especially essential where the answer has been given. It is your job to convince the marker that you know how to reach the answer. Making big jumps in your head will not do so.
- If you get stuck in the exam, perhaps even unable to see how to start your answer, write down your thoughts. There are two reasons for doing this. Firstly, what you write may be worth some credit, i.e. show what you know. You might gain the one mark you need to raise your result from a fail to grade C or even from B to A. Secondly, looking at what you have written down has more chance of leading you to an answer than looking at a blank page. Part of your strategy for becoming unstuck should include examining the words in the question to see what associations they trigger off or what equivalent words or expressions you know which might lead you to a strategy for solving the problem. The word "tangent" for example might trigger ideas about differentiating, perpendicular to the radius, equal roots, trigonometry etc. You will have used the vast majority of strategies required to complete the exam in the classroom already, so try to recall similar situations from your own experience.
- Do not start with the last question in the paper – the order of the questions has been carefully thought out, based on long experience and statistical analysis of previous exam papers, so that the majority of pupils in the country should find that the paper gets slightly harder as you work through it. This can never be totally true for every candidate, but it gives you a good guide as to which questions you might find a bit easier than others.
- Take your time with the early questions. You will feel more confident in the later parts of the paper if you know in the back of your mind that you have made a good start. Many pupils who have achieved a C pass have done so by scoring close to full marks on the first half of the paper and picking up only the odd mark here and there later on. If you are only aiming for a C pass, it might pay you to spend two thirds of the time on the first half of the paper. Remember too that sometimes the so-called harder questions at the end often require a few easy marks to lead in to the more difficult marks. So make sure you try every question, even if you can only start some of the later ones. Again, show what you know.
- Do not spend too long on any one question. There is just over a minute for each mark. If you get stuck, you may have more chance of scoring marks by trying the remaining questions first and coming back to where you were stuck.
- Remain in the exam room for all of the allotted time. Spend all of that time wisely. Look over what you have done. Have another look at anything you have had to omit. It may amaze you just how easily you can have inspiration during the last five minutes of an exam, and the marks gained then might make all the difference between passing and failing or between an A and a B pass.

- Have all the formulae and facts at your fingertips. For Paper 1, for example, it is essential that you know all the exact trig values of $0°$, $30°$, $45°$, $60°$ and $90°$ including when these angles are expressed in radians.
- If one part of a question is the form 'Show that .....' or 'Prove that........', then, even though you do not have a clue how to do this part, it is still possible for you to score full marks for any following parts.
  Consider a question with part (a) Show that $x = 30°$ and (b)..... If you cannot do part (a), you should assume $x = 30°$ and continue with part (b) just as if you had managed to prove that $x = 30°$ in part (a).
  If, of course, you try part (a) and obtain a different answer say $x = 60°$, do not use this $60°$ in part (b) but the $30°$ given in part (a).
- It is now recognised that sipping water aids concentration during exams (and in class). Check if you would be allowed to take a small non-spillable bottle of water in to your exam.

Avoiding Errors

- Arguably more marks are lost through making casual errors than not knowing how to do a question. So error avoidance is the name of the game. If you can check something by doing it another way, you may identify an error if your two answers do not agree. Where possible, check every line as you go along, e.g.
  if you factorise, expand the brackets mentally to confirm that the factors are correct;
  if you integrate, differentiate your answer to see if you retrieve the original function.
- Read each question carefully, several times if necessary, until you understand it. Make sure you are answering the question that has been asked, and not something completely different. You will not get any credit for answering the wrong question, even if it is a harder one. After completing the question, re-read it to ensure that you have left nothing out.
- Copy the information from each question down correctly. You will be penalised (usually one mark) for incorrect copying from the question paper. If your error makes the question easier, you can lose even more marks. Your error might even make the question impossible.
- Do not make invalid assumptions based on a diagram on the question paper, e.g. assuming that an angle is right because it looks like it. Diagrams are only included to give you some assistance and may not be accurate in all respects.
- Write legibly. It is remarkably common for pupils to confuse badly written 6s and 0s or badly written 7s and 1s when going from one line of working to the next. This is a self inflicted penalty. The marker must also be able to check all of your working so if it is illegible you may lose credit for what you have done.
- Be aware of 'exam vocabulary' and interpret it correctly.
  1. **show that** or **prove** mean the same thing and you are expected to justify how you can deduce the given answer, so do not omit any logical steps.
  2. **state** or **write down** means just that and will most likely only gain a single mark, so do not write an essay or attempt a proof. It also indicates that only a few steps are involved.
  3. **hence** means based on your previous working or the result you have just obtained, and is very often included to point you in the right direction, so take the advice.
- Do not begin your answer to a question near the foot of a right hand page. Turn over and give yourself plenty of room. There is less chance of making an error when turning the page, and you can view your complete solution at once, which can be useful if you get stuck later on or are looking for an error.

- Make sure you are aware of all the questions in the paper. There are always guides at the foot of a right hand page telling you to turn over. Keep going till you encounter END OF QUESTION PAPER, which always appears.

Being helpful to the marker

- Write your name neatly on the front of your script.
- Number the questions correctly and write nothing else in the left hand margin.
- Do not write anything at all in the right hand margins. Markers and checkers use those.
- Never write two questions side by side. Always write each question or part of a question underneath what goes before. Saving paper is not your concern at this time.
- If you omit a question (or part of one) and have to come back to it later but find that you have not left enough room and so offer more working further on in your answer book, make sure that you write 'see later' or something like 'turn 5 pages' beside your first attempt. The marker does not wish to award you a score for a question and then have to go and re-assess the question when the second attempt or continuation is met.
- In some of your answers there may be some doubt as to whether or not you deserve a particular mark. The marker is allowed to use discretion to award a mark on the basis of benefit of the doubt. You are more likely to score this mark if you have tried to create a good impression on the marker and made the marking as straightforward as you can.

Some Rules

- If you make two attempts at the same question, one of them must be deleted, but do not delete the first until you have written the second. Then decide carefully what you want the marker to read before you delete what you think is incorrect. Make sure you do not score out any working that is an essential part of your final solution. Correct working scored out is not given any credit. Remember that you do not get two bites at the cherry; one version must be scored through with the pen.
- All your working must be done in the answer book or on any additional pages that you are given. You are not allowed to write on scraps of paper. You are not allowed to take your exam script out of the exam hall, or tear any pages out of it. You are, however, allowed to make pencil markings on the diagrams in the printed question paper, which is yours to keep.
- Do ask an invigilator about any poor printing or suspected omissions. They can occur.
- Make absolutely no attempt to cheat. If you are caught, and you would not be the first, you jeopardise your whole certificate, not just your chances in Higher Maths. Do not do anything contentious, though perhaps innocent, such as taking out your diary or your timetable; these may be considered to contain information to which you should not have access during the exam.

And finally above all, **do not panic**. Even if you think you have made a 'fatal error', keep cool and do your best with the rest of the paper. You may have time to rectify the problem, or it might not be nearly so bad as you think.

APPENDIX TWO: NOTATION

Notation

It is essential for you to understand and use mathematical notation and conventions. Here are some general symbols and conventions used throughout this book and elsewhere. Specific topic related notations, e.g. for vectors, are explained within the text.

Numbers

- numbers with 5 digits or more are grouped in threes from the point, e.g. 52 563 0·005 21
- separate the groups with half spaces, not commas
- the decimal point appears 'half way up': 4·5 not 4.5
  (4.5 is an alternative notation for 4×5, commonly used in number theory)

Logic

- $p \Rightarrow q$ (read as p implies q) means that if the statement p is true, then so is q.
  There should really always be a connective between successive lines of working, e.g. in the solution of an equation $3(x+2) = 5(x-1) \Rightarrow 3x + 6 = 5x - 5$
  $\Rightarrow$ can also be replaced by $\therefore$ (therefore) but NOT by another = sign
- $\because$ can be used for since (or because)
- $\Leftrightarrow$ means if and only if i.e. $p \Leftrightarrow q$ means $p \Rightarrow q$ AND $q \Rightarrow p$

Sets

- $x \in \mathbb{R}$ means that x is a member of $\mathbb{R}$, the set of real numbers
- $A \cap B$ denotes the intersection of the sets A and B, i.e. the elements that belong to both A and B, e.g. the intersection of the set of prime numbers and the set of even numbers is {2}
- $A \cup B$ denotes the union of the sets A and B, i.e. the set of elements that belong to A or B (or both), e.g. the union of the set of rational numbers and the set of irrational numbers is $\mathbb{R}$ (This is particularly useful when writing the solution set of some quadratic inequalities, e.g. example **17·8**)

Functions

- when we write $f(x) = 2x + 3$, the name of the function is f; $f(x)$ shows that f is a function of x
- when we write sin ($x°$) in preference to $\sin x°$, this reinforces the fact that 'sin' is a function. Some calculators now open up the bracket automatically when you type sin.
- I have not been consistent in including these brackets every time. Your teacher may not believe that they are necessary. You should be aware of both notations.
- Including these brackets should reduce the chances of making horrendous mistakes of the kind where $\frac{\sin x}{\sin y}$ is cancelled down to give $\frac{x}{y}$.

APPENDIX THREE: MATHEMATICAL VOCABULARY AND INDEX

You do not wish to be precluded from scoring the marks for a question because you do not understand the language that has been used in the question. Many mathematical words simply need to be learned. You should be familiar with every word and phrase in the list. It doubles as an index to the book because I have given at least one item number where you can find each key word in the list.

algebraic 2·7
altitude 3·15
amplitude 7·10
angle between a line and a plane 7·4
angle between two planes 7·4
angle between two vectors 13·12
area between two curves 10·7
area under a curve 10·6
ascending order 9·1
asymptote 2·6
auxiliary angle 16·2
base (of a logarithm) 15·3
basis vector 13·11
behaviour for large x 5·10
Cartesian diagram 3·12
chain rule 14·2
codomain 4·1
coefficient 2·1
coincident 12·8
collinear 3·7
column vector 13·5
completing the square 4·9
component 13·1
composite function 4·4
compound angle 11·1
concurrent 3·13
constant of integration 10·2
cubic 9·5
curve sketching 5·10
decreasing 5·8
definite integral 10·5
degree 9·1
derivative 5·1
derived function 5·1
descending order 9·1
difference of two squares 2·4
differentiate 5·1
discriminant 8·3
distributive law 13·13
domain 4·1
exact value 7·2
exponential 2·6
function 4·1
function of a function 14·2
general equation of a circle 12·3
gradient 3·3
identity 7·1
image 4·1
increasing 5·8
indefinite integral 10·2
index (indices) 2·2
inflection 5·9
integrand 18·4
integrate 10·1
intercept 3·8
intersection 3·11
inverse function 4·5
inverse process 10·1
irrational 8·3
justify 6·3
limit 6·3
limits 10·2
linear 2·1
linear relationship 15·7
locus 3·10
logarithm 4·7
magnitude 13·1
maximum 4·9
median 3·14
minimum 4·9
multiple angle 11·1
nature table 5·9
n^{th} term 6·1
one to one correspondence 4·2
optimization 5·12
parabola 2·6
parallelogram law 13·2
period 7·10
perfect square 2·3
perpendicular bisector 3·16
polynomial 9·1
position vector 13·1
power 9·1
projection 7·4
quadratic equation 2·5
quadratic formula 2·5
quadratic function 4·8
quadratic inequality 8·1
quartic 9·5
quotient 5·4
radian 7·5
range 4·1
rate of change 5·1
rational 8·3
rational denominator 8·6
recurrence relation 6·1
reflection 4·6
rotation 4·8
scalar 13·1
scalar multiple 13·2
scalar product 13·10
second derivative 5·9
section formula 13·9
sense 13·1
stationary 5·9
surd 2·3
synthetic division 9·2
tangent 12·5
translation 4·8
triangle law 13·2
union 8·4
unit vector 13·5
unreal 8·3
vector 13·1
zeros 7·11